PICTURES, POP BOTTLES AND PILLS

PICTURES, POP BOTTLES AND PILLS

Kodak Electronics Technology That Made a Better World But Didn't Save the Day

K. Bradley Paxton, Ph.D.

Fossil Press, Rochester, New York

Pictures, Pop Bottles and Pills:
  Kodak Electronics Technology That Made a Better World But Didn't Save the Day

by K. Bradley Paxton, Ph.D.

Copyright © 2013 by K. Bradley Paxton, Ph.D.

All rights reserved. No part of this book may be reproduced or transmitted in any form or by any mechanical or electronic means, including photocopy, recording, or any information storage and retrieval system, without written permission of the author, except in the case of brief quotations embodied in critical articles and reviews.

Requests for permission to make copies of any part of this publication should be mailed to the author at ADI, LLC, 855 Rolins Run, Webster, NY 14580.

Published by Fossil Press
100 Parkwood Ave.
Rochester, NY 14620–3404

ISBN 978-0-9910216-0-4

Printed in the United States of America

Cover photo and design by Frank Cost

To Joyce, Ken, and Holly,

who lived it with me

CONTENTS

	FOREWORD	xiii
	1 WHY THIS BOOK?	1
	2 THE SIXTIES	
1960	Recordak Reliant Microfilmer	4
	IBM 1620 Computer	6
1961	Carousel Projector	8
1962	Early Development of Electrophotography	9
1963	Pageant Sound Projector	10
	R&E at Kodak's Apparatus & Optical Division	12
	Gambit Spy Satellites ... shhh!	12
	Instamatic Camera	14
1965	Super-8	15
1966	Lunar Orbiter Project	15
	Lunar Mapping & Survey System (LM&SS)	20
1967	Off to Get my Ph.D.	21
	Elmgrove Plant	21
1969	Kodak Colorado Plant	22
	Apollo Stereo Camera	22
	End of the Sixties	22
	3 THE SEVENTIES	
1972	Pocket Instamatic Camera	23
1973	Spin Physics	25
1974	Start of the Disc Camera Program	25
1975	Ektaprint Copier-Duplicator	26

3 THE SEVENTIES, contd.

Year	Topic	Page
1975	Electronic Still Camera Prototype	29
	Bayer Color Filter Array	34
	Supermatic Film Video Player	35
1976	Instant Camera	36
	Sony Betamax VCR	37
	Kodak's Secret VCR Project	37
1978	Eastman KodaPak Thermoplastic Polyester	38
	End of the Seventies	38

4 THE EIGHTIES

Year	Topic	Page
1980	Ektachem Analyzer	40
	Spin Physics Motion Analysis System	41
1981	IBM PC Model 5150	41
	SONY MAVICA	42
1982	Disc Photography	43
1983	Mead Digital / Diconix / Scitex	45
	Consumer Products Development	46
1984	New Business Units	46
	Diconix Portable Inkjet Printer	47
	Kodavision Camcorder	47
	Apple Macintosh Computer	49
1985	A Hybrid Approach	49
	KEEPS and KIMS	50
	Color Video Imager	50
	Modular Video System	52
	Eikonix Designmaster	52
1986	UltraLife Lithium Power Cells	53
	World's First Megapixel CCD Sensor	53
	Demise of Kodavision	54
	Electronic Photography Division	55
	Still Video Camera Design	57
	UltraTech	59
	Ektaprint Digital Printer	59

4 THE EIGHTIES, contd.

1986	Ektascan Medical Laser Printer	59
	Verbatim Optical Disc	60
1987	Video ID System	60
	Still Video System	62
	Fling Single-Use Camera	68
	Organic Light-Emitting Diodes	69
1988	Sterling Drug Acquisition	69
	ColorEdge Copier	70
	Qualex	70
	Polaroid Damage Settlement	71
	What to Do with Diconix?	71
	Create-A-Print	71
	Slide/Video Transfer Unit	72
	Prism XLC Electronic Previewing System	72
	Kodak/Hitachi Technology Agreement	74
	Tactical Camera	74
1989	Digital Continuous Tone Printer	76
	More One-Time-Use Cameras	77
	End of the Eighties	77

5 THE NINETIES

1990	Photo CD System	78
	Adobe Photoshop	81
	Hubble Trouble	82
1991	Professional Digital Camera System	83
	Approval Proofing System	84
	Technology Drivers	85
	A Valure Model for Cameras	87
1992	The Big Buyout	87
	More New Digital Products	88
1993	Portable Photo CD Player	88
	Cineon	88
	Imagelink Scanner	89

5 THE NINETIES, contd.

Year	Entry	Page
1993	Diconix Sold	90
	Kodak Picture Exchange	90
	ColorEase Printer	91
	Eastman Chemical Spun Off	91
1994	Sterling Drug Sold	92
	Associated Press Photojournalist Camera	92
	Apple Quicktake Camera	93
	Picture Maker Kiosk	94
	Fuji Sponsors the Olympics	94
1995	Point-and-Shoot Digital Camera	94
	Copier Sales and Service by Danka	95
1996	Spirit DataCine	96
	Advanced Photo System	96
1997	Danka Acquires Kodak's Copier Business	98
	Kodak Picture Network	99
	Digital Science Zoom Digital Camera	99
	Wang Software Acquired	99
1998	You've Got Pictures!	100
	Kodak Professional DCS Cameras	100
1999	Copier Manufacturing Sold to Heidelberg	100
	First OLED Display	101
	End of the Nineties	101

6 THE TWO THOUSANDS

Year	Entry	Page
2000	Phogenix Joint Venture with Hewlett-Packard	102
2001	Bell & Howell Imaging Acquired	102
	EasyShare System	102
	Kodak Acquires Ofoto	102
2002	OLED Prototype Display Shown	103
	Encad Large Format Printer	103
	End of the Kodak Professional Camera Business	104
2003	EasyShare Printer Dock	104
	EasyShare Zoom Digital Camera	105

6 THE TWO THOUSANDS, contd.

Year	Entry	Page
2004	Scitex Repurchased by Kodak	106
	NexPress Digital Production Color Press	106
	Remote Sensing Systems Business Sold to ITT	106
	Ultralife Sold	107
	Kodak Out of Camera Business!	107
2005	Kodak Polychrome Graphics	107
	What's in a Name?	108
	CMOS Devices Announced	108
2006	More Digital Cameras	108
	New Kodak Logo	108
2007	EasyShare All-in-One Printer	108
	Health Business Sold to Onex	109
	EasyShare Digital Camera	109
	U.S. Film and Digital Camera Sales	110
2008	Mobile Phone Sensor	111
	Stream Inkjet Technology	111
	First CCD with 50 Megapixels!	111
	Wireless All-in-One Printer	112
2009	Stream and Prosper	112
	Pocket Video Cameras	112
	Bell & Howell Scanners Acquired	112
	Kodak Sells OLED Technology	112
	End of the 2000s	113

7 THE TWO THOUSAND TENS

Year	Entry	Page
2010	Prosper Press	115
	Another All-in-One Inkjet Printer	115
2011	Selling More Businesses	116
	Scanmate Scanner	116
	Versamark Printing Systems	116
	Laser Projection Technology	117
	Kodak Spy Satellite Programs De-Classified After 50 Years	117
2012	Kodak Declares Bankruptcy	124

7 THE TWO THOUSAND TENS, contd.

2012	Prosper Presses Expanded	124
	Kodak is Out of the Digital Camera Business	124
	Kodak Gallery Sold to Shutterfly	125
	ITT Buys Space Computer Corp.	125
	Kodak Tries to Sell Patents	125
	Kodak Sells More Businesses	125
	Kodak Quits Making Desktop Inkjet Printers	126
	Kodak Sells Digital Imaging Patents	126
2013	Innovations Since 2012	126

8 SO WHY DIDN'T IT WORK?

What Business Are You In?	128
Constancy of Purpose	129
What Is the Picture Business, Anyway?	130
Four Conclusions	133
Personal Postscript	135

ENDNOTES	137
REFERENCES	139
ILLUSTRATION CREDITS	141
ACKNOWLEDGEMENTS	143

FOREWORD

In my first job at Kodak in 1965, I was ushered into a four-person office that had one empty desk. My manager, Jim Maher, told me, "You have big shoes to fill when you sit at that desk. That was Brad Paxton's desk." As time went on I was to find out how right Jim was. At that time Brad and I were both engaged in one of the most exciting experiences of my career. We were part of a team that was creating a photographic spy satellite.

As our careers progressed, our paths crossed many times until we both finally retired from Kodak in 1992. By then I was a senior vice president. I subsequently joined the University of Rochester's Simon School of Business Administration faculty as an executive professor, and 21 years later, I am still teaching. Brad and I remain close friends. He retired as director of the Electronic Imaging Research Labs. In that role, Brad guided the development of a wide range of Kodak electronic imaging products and systems. This book takes us from Brad's experience in 1960 as a young engineer developing Kodak's first transistorized electronic product, a sound amplifier for a sound projector, to the advanced electronic cameras and Internet systems of the 21st century.

Kodak was a large firm involved for many years in government, commercial and consumer markets. It had a world class R&D capability. Much has already been written about Kodak's prowess in photographic chemical research. This book provides a unique perspective on the equipment side of the business. It focuses on Kodak's use of electronics in developing innovative products for a wide variety of applications.

But this is not a technical book about electrons and photons. Rather, it is about people and their innovative ideas, and how these ideas were brought to life in new products. It is about the challenges of moving ideas from a development lab to the marketplace, and how some management decisions influenced that movement.

Brad's candid, first-hand account of his career at Kodak explores two main story lines. The first is about the evolution of electronics at Kodak and how it expanded with the explosion in electronics technology. Brad includes Kodak's spy satellites, photographing the moon from a moon-orbiting satellite, office copiers, automated document management, and leading-edge electronic products in consumer and commercial markets, both those that were successful and some that never made it to market. Brad's breadth of experience across Kodak's diverse markets makes this a fascinating story.

The book's second story line is Brad's personal reflection of how Kodak went from being a corporate giant to bankruptcy in 20 years. This is a no-holds-barred evaluation of what happened, why it happened, and perhaps how it might have been avoided. Brad's personal view will not be found in newspaper editorial pages or in clever Wall Street analyses.

As a well-respected Kodak vice president, Brad lived in the midst of Kodak's efforts to succeed in the inevitable transformation from traditional photography to digital photography. His department was charged with using electronics to innovate and create great products, and yet management, I suggest, was reluctant to commercialize the electronic innovations they made, realizing that these new products would ultimately replace Kodak's traditional photographic products.

Some readers may want to understand what Kodak was up to during this transition to digital photography, and whether it was possible that such a great company was asleep at the switch. Others who knew Kodak well, like current and former employees, customers, suppliers, and investors, will enjoy learning more about the Kodak products and management decisions that affected their lives. These readers will be very interested in Brad's pithy analysis of what happened at Kodak.

I thought I knew what was going on within Kodak when I was there, but I learned a lot from reading this book. Brad discusses many interesting Kodak projects, such as:

- **The development of the spy satellite.** Developed in the 1960s, it was kept secret until finally being declassified after 50 years in 2011. This program was tremendously successful and a vital national defense intelligence asset at the height of the Cold War.
- **The *Lunar Orbiter* project.** It produced a satellite that orbited the moon, photographically mapped its entire surface, and then transmitted the images electronically from the moon back to NASA. These images provided NASA the information it needed to determine where astronauts could land.
- **The development of a prototype of an electronic camera in the 1970s.** This has been in the news frequently since Kodak's bankruptcy last year, because Kodak had been erroneously accused of missing the electronic camera innovation. This prototype not only demonstrated that Kodak was not asleep but it is now understood to be the first digital electronic camera. Once again, Brad was in the midst of the action, and the book provides little-known details about this development.
- **Kodak's entrance into the commercial electrophotographic market with its *Ektaprint* copiers in the 1970s.** The *Ektaprint* immediately became the market standard of quality. Brad's innovative image-chain analysis of the *Ektaprint* copier provided the roadmap that guided developers to *Ektaprint* quality.
- **The mid-1980s development of an electro-optical camera.** When one of Kodak's government customers expressed an interest in a lab prototype of this camera, we built it, the agency took it for evaluation, and we never saw it again. A refined prototype called the *Tactical Camera* was then developed and two were manufactured, one of which has been found and given to George Eastman House in Rochester, NY. Todd Gustavson, curator of technology at George Eastman House, commented, "The *Tactical Camera* may well be the most important object acquired during my 24 years at Eastman House. There is nothing like it in the collection, but more importantly, it is one of only two of its kind ever made, and it is from these models that all digital cameras were derived."

Over the following years, Kodak's Electronic Photography Division and others developed an accelerating stream of electronic imaging and systems. Even though they were all innovative and provided real consumer value, some of them never made it to market. Brad not only discusses the ones that were commercialized but also, and perhaps more fascinating, a number of leading-edge products that were detoured on the way to market and remained in the lab.

When Brad's Kodak career began, the company was a monopoly (although we were not allowed to say or write that word in the 1960s). For the next three decades, Kodak increasingly employed electronics in its various business units, and Brad was a leader in many of these efforts. His Kodak experience uniquely qualifies him to tell this story about the people who used electronics to innovate and create great products, about management's understandable fear of giving up extraordinary profits from its traditional photography products, and about the cost of Kodak's losing its constancy of purpose.

Lawrence J. Matteson
Rochester, New York
August 2013

PICTURES, POP BOTTLES AND PILLS

Kodak Electronics Technology That Made a Better World But Didn't Save the Day

1

WHY THIS BOOK?

Many economists and Kodak retirees followed the story of Kodak's slow decline over the past twenty years with fascination. My long-time friend Dr. Joseph DiStefano of the Institute for Management Development (IMD) in Switzerland had been bugging me since the mid-1990s to write "something about Kodak." I always rebuffed him with, "I'm waiting to see how it all shakes out," or later, "I don't know the title yet." Those responses worked for a long time, and then Kodak filed for Chapter 11 Bankruptcy on January 19, 2012. Now I was out of excuses.

My previous boss and colleague, Professor Larry Matteson at the Simon School of Business at the University of Rochester (U of R), thought that the bankruptcy announcement was a very sad turn of events, and we both hoped that the Simon School would write a completely objective business case on Kodak, not biased in any way. I became interested in the project because I had had a great career there from 1960 to 1992. Although there were no doubt plenty of people who knew more about finance, management, culture, marketing and other aspects of the company, I could write with some authority on electronics and equipment technology development at Kodak because I was there for 32 years and was heavily involved in a lot of it.

Kodak's Apparatus and Optical Division (A&OD) was formed in 1956 to develop, design and manufacture photographic equipment. It included two main sites in Rochester, NY: Camera Works and the Hawkeye Works. I started working in the electrical lab at Kodak's Camera Works as a recently-graduated electrical engineer from Rensselaer Polytechnic Institute. Kodak hired me to design a sound amplifier made of "these new things called transistors" for the *Pageant 16 mm Sound Projector,* which at the time had a (rather heavy) tube-type amplifier.

During my subsequent enjoyable career at Kodak, I worked almost exclusively on electronics-related undertakings. I was involved in the *Lunar Orbiter* program, secret space reconnaissance programs, electrophotographic copiers, various cameras, many printers, photo kiosks, video imaging, and digital imaging. Even though I never was personally especially ambitious, management-wise, I became the general manager and vice president of Kodak's Electronic Photography Division and subsequently of the Printer Products Division. At the end of my career I wound up managing Kodak's electronic imaging research labs along with Jim Meyer, who ran the traditional imaging research labs, and we both reported to Ed Przybylowitz, who was Kodak's chief technical officer.

After 32 years at Kodak, I left in early 1992, not because I was ready to retire, but because Kodak was tempting middle managers to take an early leave under a "3R" plan (Resource, Redeployment, and Retirement). I didn't know the precise personnel numbers at the time, but learned later on that Kodak was looking to eliminate up to 3,000 positions, and the offer was so juicy that more like 8,000 employees actually took it, including me. My wife, Joyce, made the clarifying observation, "You can do anything you want," which I somehow found profoundly compelling.

I started my own consulting company in Rochester in 1992, KBPaxton, Inc., and began working out of the now-defunct Rochester Institute of Technology (RIT) Research Corporation. We began to assist the U.S. Census Bureau in 1993, helping it to use digital electronic imaging to process the 2000 Decennial Census. Advanced Document Imaging (ADI, LLC) was formed in 2002, and we continue to help the Bureau to this day, as well as many other clients interested in forms processing systems and information technology (IT) classification systems. My first book, *Handprint Data Capture in Forms Processing: A Systems Approach*, gives a thorough overview of that 19-year census experience, although it's a tad technical, and, as a census colleague told me, "probably not a *New York Times* best-seller." Looking back, I was perhaps more successful at encouraging the U.S. Census Bureau to use electronics than steering Kodak in that direction.

For this study of Kodak, I decided to take a close look at the various electronic technologies, equipment products, and systems that the company developed from 1960 to the present, and then analyze this rather large pile of data to see whether it would tell me something factual and instructive about what had transpired that was not already obvious. My results became this book. Whether I have shed some useful light on the tragic story of a significant world-famous company brought to its knees will be for others who follow to judge.

My approach was to take selected product and technology cases by year, grouped by decade, and trace what was done in each case, how significant it was to the world at large or to Kodak's business, and then to determine its eventual outcome. This chronological scheme is taken year-by-year in a kind of annotated timeline. My overall conclusions are presented in the last chapter, and I am available to discuss or receive comments at brad.paxton@adillc.net.

2

THE SIXTIES

Transistors were just coming into practical usefulness when I started at Kodak, and this seemed to me to be as good a place as any to start my study for this book. Elaborate treatments of traditional Kodak cameras, chemicals, and films have been worked over pretty well by now, going back to 1888, when George Eastman exclaimed, "You press the button, we do the rest." In this study I am intentionally leaving out film developments per se, except as a part of certain (hybrid) imaging systems. My focus is electronics.

Starting out in Kodak's Camera Works, working on the transistor amplifier for the Pageant projectors, I met a chap named Earle Young. Earle ran an engineering group at Kodak's Hawkeye Plant in Rochester, mostly developing microfilm equipment. When he found out that I had just been hired, he kindly offered me a tour of his department. As a result, one of the first equipment products I became familiar with (besides the ones I was working on) was Kodak's microfilm equipment, famously sold under the trade name *Recordak*.

1960
Recordak Reliant Microfilmer
The *Recordak Reliant 500 Microfilmer* was capable of photographing up to 500 checks or 185 letters in one minute. *Recordak* equipment was **the** way that banks and many other businesses archived their images reliably and in a small space. It is not widely known, but the paper transport in this product became the basis for the reliable paper transports of the *Imagelink* scanners to come later, further along in the electronic age.

Perhaps not fully appreciated by Kodak management at the time was the fact that reliable paper transports are not easy to come by, and that this technology was to become a "core competency"[1] that was drawn on again and again in later years. For example, while I was working on the *Ektaprint* copier in the early 1970s (see Chapter 3), Matt Russell, who designed its recirculating feeder, taught me, "Good paper paths should be short and straight." Later on, one of the main reasons that the Census Bureau decided to use truckloads of Kodak *Imagelink* scanners for the 2000 Census was that the paper transport system on those scanners, being a direct descendant of the system used in the *Recordak* machine, had "more miles of paper run through it than any scanner on the planet," as I used to say. (The notion of "core competency" will continue to arise throughout this story.)

Fig. 1960.1 *Recordak Reliant 500 Microfilmer* (Fortune)

The studious gentleman above is contemplating the wonders of a roll of 16 mm microfilm, highlighted with an asterisk (*) so you won't miss it. These wonders include the ability to store over 10,000 business letters on a single roll of microfilm, like the one he's holding in his hand, and other valuable attributes as outlined in the rather well-written ad shown on the next page:

These were the first microfilm readers to have index codes marked on the edges of the film. If you knew the code for the particular document you

WEIGH IT ...5 OZS. COMPARED WITH 107 POUNDS

RECORDAK microfilming takes the bulk out of office filing. Lets you keep as many as 10,500 letters on a single roll of 16mm film. You can file this tiny roll in just 1% of the space needed now. Find any record in seconds because today's RECORDAK RELIANT 500 Microfilmers can index the film with code lines which speed retrieval in Recordak Film Readers. Want a paper print of the microfilm? You get that in seconds, too! Costs are another surprise. Imagine, twenty letters on a cent's worth of film including processing charge. Call a Recordak Systems Man today for the whole exciting story. Recordak Corporation, Dept. A-4, 770 Broadway, New York 3, N.Y.

RECORDAK CORPORATION
770 Broadway, Dept. A-4, New York 3, N.Y.
Send more information.

Name_____
Company_____
Street_____
City_____Zone____State____

RECORDAK®
(Subsidiary of Eastman Kodak Company)
**first and foremost
in microfilming since 1928**
IN CANADA—Recordak of Canada Ltd., Toronto

Fig. 1960.2 *Reliant 500 Ad Text* (Fortune)

wanted, you typed it in, and the reader would go "whoosh," fast-forwarding to just past your document, and then would slowly back up and precisely register your page. It was very amusing to watch this happen on the optical viewing screen, so the time spent waiting wasn't boring. Then you pushed a button, and a print came out. It was a very cool product, and was sold worldwide.

I'm inclined to apologize for the sexist phrase "Recordak Systems Man" in the above text. However, when I started at Camera Works as a young engineer, I was told to wear a white shirt and tie. A colleague dared to wear a pink shirt one day, of all things, and was sent home. I guess you had to be there. In 1960.

IBM 1620 Computer

Another important development from 1960 was not originated at Kodak, but rather at IBM, the *IBM 1620 Computer*. I will occasionally toss in one of

Fig. 1960.3 *IBM 1620 Computer (Todd Dailey/Wiki)*

these "external to Kodak" electronic developments when it provides some context for Kodak electronic developments that are the primary focus here. Actually, it is important to note that Kodak was not the *only* company developing electronics products that became integral to the imaging market.

It didn't take long for my division at Kodak to buy one of these things, and although I had taken a theoretical computer course at RPI, this was my first serious digital computer experience. It used a form of resistor-transistor logic (RTL) employing so-called drift transistors, and had a ferrite-core memory available in 20 kilobyte (KB) chunks, housed in a steel rack taller than most of us humans. It first used paper tape and soon after used punch cards for input/output. We programmed it in FORTRAN, and thought it was great! Remember these punch cards?

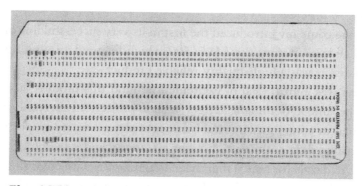

Fig. 1960.4 *IBM Punch Card (KBP)*

Marv Harrison, a brilliant aeronautical engineer/mathematician, and I were trying to invert a "stiff" matrix on the *1620* for a photo satellite optical mirror-mount analysis, and in the process, exceeded the 20 KB ferrite-core memory in an over-the-weekend computer run. As a result, due to the project's urgency (more about the satellite projects later), Kodak quickly bought another 20 KB memory, so we now had a grand total of 40 KB. We thought we were really in hog heaven then, and we successfully inverted the matrix! (I am still astonished that the iPhone in my pocket has over six gigabytes [GB] of memory, which is about 150,000 times more memory than that *IBM 1620*'s two-rack 40 KB computer had.) So now, we electrical engineers at Kodak were beginning to seriously use computers for analysis, and we were also becoming increasingly fascinated with transistors.

Fig. 1961.1 *Carousel Slide Projector (GEH)*

1961

Carousel Projector

In 1961, the company introduced the first in its very successful line of Kodak Carousel slide projectors, the *Model 500*, which featured a round tray holding 80 slides. It was a high-quality, reliable product.

There was a great deal of excitement around this product launch. The *Carousel* became a "dominant design" in slide projection; everyone speaking at a seminar or conference inevitably brought along a trusty round tray with slides. Serious slide enthusiasts bought a *Carousel* and a bunch of trays for home use to dazzle (or bore?) their friends and family with color slide shows. (The term "dominant design" will be popping up again as we go along.)

Fig. 1961.2 *Carousel "As Dependable as Gravity" Ad (GEH)*

Still at the Camera Works division, I became acquainted with the head of design, a sharp, energetic chap named Vernon (Vern) Jungjohann. Vern himself explained (as best he could to a "sparkie", i.e., an electrical engineer) how the gravity-feed mechanism worked on the *Model 500*, and furthermore, how it worked reliably every time. Note that the ad on the previous page is captioned, "As dependable as gravity."

Manufacturing the plastic molding technology that was used to make the slide tray, Vern told me, was a significant challenge due to the tight tolerances, but the experienced German machine shop tool-and-die experts at Camera Works did it! They enhanced the product later with a 140-slide tray that was even more difficult to make, because that was a feature that customers wanted. Millions of these projectors were sold worldwide in the over forty years they were on the market, and production was discontinued in 2004 due to digital projectors becoming popular.

1962

Early Development of Electrophotography

Before 1960, Kodak offered a product called the *Verifax*, a semi-dry copier that was based on silver halide technology and was sold worldwide. A master was exposed and dipped in some processing solution, and after a plain sheet of paper was placed next to the master and sent through rollers, the peeled-off paper held a (slightly damp) copy of the original image. It was very manual, and secretaries didn't like changing the processing solution because it was messy and could stain their clothing. When Xerox introduced the *Xerox 914* copier in 1959, it became clear to Kodak that the days of the *Verifax* were numbered. I saw a demonstration of a copier prototype built by Howard Hodges at Camera Works of an automatic *Verifax* process that eliminated all manual handling and produced almost dry copies, but it was never manufactured as a Kodak product. I always wondered what might have happened if Kodak had introduced that product before the *Xerox 914* came out.

Many stories have been told about how Chester Carlson, "the inventor of electrophotography," attempted to interest Kodak and others in his technology in the 1940s, all to no avail. Kodak's lack of interest was not just because the images were lousy compared to film (which they actually were back then) but also because Kodak was deeply engaged in the War effort.

After the *Xerox 914* came on the market, the idea that electrophotography could produce increasingly better images became apparent, and Kodak began to improve the image quality of this complex seven-step process (charge, expose, develop, transfer, fuse, clean, erase). One innovation that looked promising was a design using an organic photoconductor (OPC) on a belt, rather than a selenium photoconductor on a drum, the nominal and expensive Xerox approach. Dale Smith of the Kodak Research Labs (KRL) built a prototype development process using an OPC in 1962, shown here:

Fig. 1962.1 *Dale Smith's Early OPC Prototype (EKC)*

When I got involved with electrophotography in the 1970s, I met Dale, a fine scientist, a gentleman and a mountain climber in his spare time. When he was asked why he climbed mountains, Dale would say, "Because they're in my way." He was just as determined in his research. This technology continued to be successfully developed and was the heart of the *Ektaprint 100 Copier-Duplicator* that came out in 1975 and shocked Xerox. Electrophotography at Kodak became another core competency, and later evolved into Kodak's *NexPress* high-speed color printer.

1963
Pageant Sound Projector
Kodak shipped the new transistorized *Pageant Sound Projector* in 1963. These sound projectors read a film's optical sound track with a solar cell and

Fig. 1963.1 *Pageant Transistorized Sound Projector (James Smock/TSS)*

transistor-type amplifier. The earlier *Pageant* projectors, as I mentioned previously, dating from the 1950s, had heavier tube-type amplifiers. I designed the 25-watt transistor amplifier for this product; it was the primary reason Kodak hired me in 1960. It actually involved electronic imaging, too, because the sound track was recorded optically, in a wiggly variable-width image along the film edge, and was read with a photodiode illuminated by a lamp. Of course, to be heard by the audience, the signal had to be electronically amplified.

It was a terrific first electrical engineering project, as far as I was concerned, because I got to design it, put it into production, and deal with quality control and manufacturing issues along the way. I developed a lot of respect for manufacturing and QC people, and their ability to define and solve problems. It was the first Kodak commercial product that used transistors, as far as I know. It remained in production for over twenty years, and was used widely in schools and businesses to show sound movies. If you are over fifty years old, you are sure to have watched movies on one of these projectors; it had a cover (containing the speaker) that came off and, placed up near the screen, provided the sound—remember?

Research & Engineering at Kodak's Apparatus & Optical Division

After looking around Kodak for my next opportunity, I was hired by the research and engineering (R&E) organization, part of Kodak's Apparatus and Optical Division (A&OD). It turned out they were doing secret space satellite camera systems for the government, although I didn't know anything about that until after I got my secret clearance. During my interview, they described the emerging *Lunar Orbiter* project as an example of what they did, and I got to work on that also.

I worked with some astonishingly smart, talented people in R&E, including Chuck Spoelhof (who helped fix the *Hubble* telescope in the early 1990s and was named a Pioneer of National Reconnaissance by the National Reconnaissance Office in 2000), Don Smith (my math mentor), Marv Harrison, Jim Maher, Ed Warnick, Frank Hicks, Bill Feldman and many, many more. They encouraged me to start my M.S. in Applied Mathematics at the U of R evening school, which I did, and fortunately, Kodak paid the tuition for part-time graduate school in those days. The R&E group was a hotbed of technical expertise at Kodak, and many of its alumni went on to successful careers in other areas of the company, being highly trained in the core competencies of advanced imaging and project management.

Gambit Spy Satellites ... shhh!

The Cold War was in full swing, and with the encouragement of President Eisenhower, the first *Gambit 1* spy satellite was launched in 1963. The *Gambit* program flew 54 missions over twenty years, with significant improvements along the way. It feels very strange for me to even mention this event that happened in 1963, because, apart from those who were cleared for it, nobody at Kodak or anywhere else knew about the program at that time. The program was not declassified until 2011, on the occasion of the 50th anniversary of the National Reconnaissance Office (NRO).

Kodak's extremely important innovations and participation in space photography for the U.S. government is worthy of a book by itself. This work helped keep the Cold War from escalating into WWIII by informing the U.S. government about key intelligence targets in the "area of interest," as the Soviet Union was covertly referred to at the time. The dedicated men

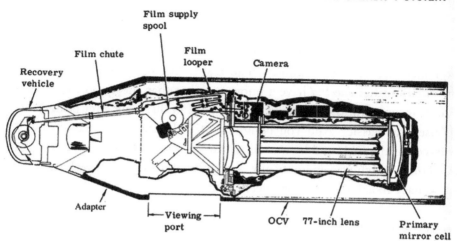

Fig. 1963.2. *Gambit 1 Reconnaissance Satellite KH-7 (AF)*

and women in the Rochester area who worked on those projects, although dwindling in number, still gather together once a month for lunch to recall the challenging but satisfying days when they were making Kodak's "eyes in the sky," and to discuss current events. More interesting details about these super high-tech programs are given in Chapter 7. However, shown above is a side-view drawing of the *Gambit 1* payload just for flavor.

Note that the size of this camera payload was rather large: 5 feet in diameter and 15 feet long, definitely bigger and a lot more complex than the *Instamatics* discussed in the next section! It would take many pages to describe all the components of the *Gambit 1*, and it was a monumental undertaking in all respects: optics, film technology, electronics, manufacturing, quality control, thermal control, orbital mechanics, systems analysis, and more, all done under a super-secret "black" program structure appropriately referred to as Key Hole (KH). The *Gambit 1* satellite was designated as KH-7. I worked on a number of these projects, especially an improvement to *Gambit*, called *Gambit-Cubed*, designated as KH-8.

In the mid-1960s, soon after being married, I went on a clandestine mission to Vandenberg Air Force Base in California to assist in monitoring the telemetry from the camera payload for a *Gambit* mission. I couldn't tell my wife, Joyce, where I was going or what for, as the entire space project at Kodak was highly classified. When she went to her usual Sunday afternoon dinner at her parents' house, she was asked, "Where's Brad?" She explained that she didn't know where I was, and also offered the helpful information that she *did* have a phone number that she could call; a lady would answer, and then I would call her back in due time. Her father's response was classic: "And you bought that?" He always suspected that I worked for the CIA, and he was close; Kodak was actually working for the Air Force, which was energetically competing with the CIA back then for supremacy in aerial reconnaissance.

Instamatic Camera

Meanwhile, back on the home front, the famous line of Kodak *Instamatic* cameras was introduced in 1963, featuring new, easy-to-use 126-size cartridge-loading film, which eventually brought amateur photography to new heights of popularity. The most basic model was the *Instamatic 100*, shown below:

Fig. 1963.3 *Instamatic 100 Camera (GEH)*

It was such a smash hit that more than 50 million *Instamatic* cameras were manufactured by 1970. The *Instamatic* was yet another dominant design. Clearly, at this time, Kodak remained serious about being in the picture business, as it had been since 1888. No doubt about it.

1965
Super-8

Kodak developed the *Super-8* film format and launched *Super-8* movies with cartridge-loading *Kodachrome II Film*. The frames in this format were a tad larger than the previous "regular" 8 mm movie film. *Super-8* was now the best way to take home (high-quality, but silent) movies. We took our Paxton family home movies on both flavors of 8 mm Kodak color film, and I eventually spliced many small rolls together into four seven-inch film reels that we recently transferred to DVDs. Still pictures and movies are not just ordinary household objects; they represent precious memories, and we treasure them.

1966
Lunar Orbiter Project

One of the greatest photographs of the 20th century, a close-up of the moon with the earth in the background, is in the Smithsonian Institution (and a four-foot copy is also in my office), since it was the very first photograph of earth from deep space.

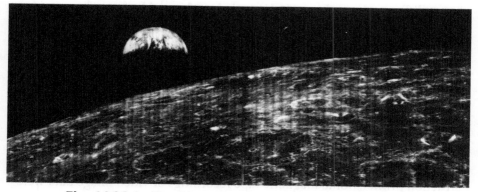

Fig. 1966.1 *First Photo of Earth from Deep Space (NASA)*

The *Lunar Orbiter 1* made this photograph, using a dual-lens camera module, film, and an on-board film processor, packaged into a compact, thermally-insulated camera payload by Kodak at the Lincoln Plant in Rochester, NY, along with a CBS electronic line scan tube called a flying-spot scanner. *Lunar Orbiter 1*'s fundamental mission was to determine safe landing areas for the NASA *Apollo Project* that came later. The film processor used a precursor to instant film processing, developed at the Kodak Research Labs in the 1950s, called "bi-mat." A sensitized strip of film was exposed and then brought into contact under pressure with another strip of developing material. The strips were then separated, and the flying-spot scanner scanned the developed film, sending radio signals representing the image down to NASA's Jet Propulsion Labs (JPL) for reconstruction on the ground, strip by strip, as you can clearly see on the photograph.

Actually, there were three Deep Space Network (DSN) stations as part of NASA/JPL that received the signals from the *Lunar Orbiter* and did the primary image reconstruction on the ground. One was at the Goldstone Deep Space Communications Complex in California, the second was in Madrid, Spain, and the third was in Woomera, Australia. (Goldstone is named for a nearby gold-mining ghost town). This seemingly odd geographic arrangement of DSN stations is still used today because these locations insure that regardless of where a satellite is relative to the earth, it is in the sight of at least one of these three stations.

When I first met my colleague Art Cosgrove in the late 1960s, he told me that his first Kodak job was video engineer at all three of these stations but at different times: *Lunar Orbiter* missions 1 and 4 in Australia, missions 2 and 5 in Spain, and mission 3 in California. Not a bad first assignment! Each station received (when it could) the radio frequency (RF) signals, demodulated them, and sent the telemetry data to magnetic tape, and the photographic data to the ground reconstruction equipment. The video images were then displayed line-by-line on a kinescope face and recorded on a continuously-moving 35 mm filmstrip. The film was chemically processed at the DSN, evaluated for quality, and physically shipped to the Lincoln Plant in Rochester, NY where the images were reassembled at the reconstruction printer. Some magnetic tape recordings of video image data were made also, but they were not considered the primary image record. The "permanent" image record was film in 1966 and 1967.

We did a lot of systems studies for this project, including a math model of what we called the V/h sensor. Here, V stands for the satellite velocity over the lunar surface, and h stands for the altitude of the spacecraft above the lunar terrain. The ratio V/h was an angular rate that could be used, along with the known focal length of the lens, to determine how fast the image was moving relative to the film plane. The film was then moved during exposure at the right speed to compensate for image motion and to get a sharp image. We also analyzed the resultant small image-to-film motions that could not be compensated, resulting in what we called "smear"; the amount of residual smear had to be carefully controlled in any photographic satellite system to get good pictures.

I was fascinated by this entire program, and was even learning something about orbital mechanics, which made me skeptical that the rocket and photo payload system would orbit the moon properly (it had never been done before), much less perform all the complex functions required once in orbit. Dr. Feldman, Kodak's *Lunar Orbiter* program manager, was intrigued and somewhat amused at a probability analysis I devised which suggested that there was only a 10% chance that the *Lunar Orbiter* would ever be able to take good photographs. Now here's a lesson in good management: Rather than chew me out, he gave a talk to the entire staff the following week about all the other system aspects of the project (many of which were unknown to us). The event boosted team morale. We more clearly realized that our complex photographic payload was but a small part of the entire *Lunar Orbiter* system. All five of the *Lunar Orbiter* missions

Fig. 1966.2. *Lunar Orbiter 4 Launch (NASA)*

were successful, the appropriate landing sites for *Apollo* were selected, and NASA went forth to make history with the first manned landing on the moon in 1969.

The photograph of the *Lunar Orbiter 4* being launched in 1966, shown on the previous page, clearly demonstrates that launching a satellite wasn't a walk in the park. The predecessor to the non-classified *Lunar Orbiter* series was a secret reconnaissance satellite called *SAMOS* (*Satellite and Missile Observation System*). Kodak also built this camera payload in the 1950s using bi-mat film. Most of the *SAMOS* missions either failed to return images or blew up on the launch pad because rocket technology was being refined at about the same time as the camera systems were being developed.

In the *Lunar Orbiter* photograph shown below, the photographic payload built by Kodak is the small roundish object in the center with two lenses looking at you: the large one for detailed imagery, and the smaller one for location and mapping.

Fig. 1966.3. *Lunar Orbiter with the Photographic Payload (NASA)*

Of course, when in orbit, the lenses would not be looking at you; they would be pointing down at the lunar surface. An artist's conception of what the *Lunar Orbiter* looked like while orbiting the moon and actually taking pictures in shown on the next page.

One of the actual *Lunar Orbiter* photographic modules from Kodak's Lincoln Plant is on display at George Eastman House in Rochester, NY. It had some transistors in it, too, which you can see at the Eastman House if you look carefully at the circuit boards on top. The module may look clunky today, but it was unquestionably at the leading edge of the super-high-tech picture business back then.

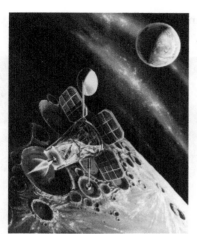

Fig. 1966.4
Artist's conception of the Lunar Orbiter at work (U of R)

Here is a view of the module being built and tested at Lincoln Plant:

Fig. 1966.5
Lunar Orbiter Payload at Lincoln Plant (U of R)

Much later, in the new century, NASA would engage in reviving the old *Lunar Orbiter* magnetic tapes, enhancing the images using new computer image processing technology. The project became known as the *Lunar Orbiter* Image Recovery Project (LOIRP), and was stimulated by my colleague and fellow space enthusiast Philip Horzempa, aided by JPL Archivist Nancy Evans, who had (luckily) refused to scrap the tapes. These enhanced images are still useful today for scientific research.

An example of this image enhancement applied to the famous earthrise photo is shown below:

Fig. 1966.6. *Lunar Orbiter Enhanced Earthrise Image (NASA)*

This Kodak photographic contribution to the NASA Apollo program is not widely known, but was very significant to our Nation.

Lunar Mapping & Survey System (LM&SS)

One more sideline to the *Lunar Orbiter* project was its sister project under NASA's Lunar Reconnaissance Program (LRP), called LM&SS (Lunar Mapping & Survey System).[2] When I worked on this, we called it the "Y" Program, operating out of the "Candy Factory," which was literally an old candy factory across the street from Kodak's Hawkeye Plant on St. Paul Street in Rochester. At Kodak, all our space programs were called by (apparently) meaningless letters to enhance security. I didn't know back then that the Y program was secretly called LM&SS and also UPWARD. The idea was to use a modified *Gambit* reconnaissance system in a manned spacecraft atop a *Saturn V* missile to orbit the moon and take pictures to aid in *Apollo* landing site selection, in case the *Lunar Orbiter* didn't work. The program was managed by Paul Murfin, a Kodak/*SAMOS* alumnus from the 1950s. On this project I now had my first systems group, and we were doing some very interesting system studies involving image motion compensation, lens modulation transfer function, and orbital mechanics.

This was a great example of the NRO cooperating with NASA even though the NRO was still a secret organization. The project came to a sudden end in the summer of 1967, largely because the *Lunar Orbiter* worked, and very well at that. It seemed like a good time for me to get involved in my next adventure—the University of Rochester.

1967

Off to Get my Ph.D.

My personal experience with the excellent Kodak people working in R&E on the space programs encouraged me, after completing my M.S. in Applied Math, to continue my education. So I took a leave of absence in the fall of 1967 to get a Ph.D. in electrical engineering at the University of Rochester, not really being sure how the Paxtons would survive the adventure. Fortunately, I was given a Kodak Doctoral Award that supported my family and me for two years. I mention this only because to me, this was a concrete example of a great company really caring about developing its people, and looking long-term into the future.

Elmgrove Plant

During this time, the Camera Works plant on State Street in Rochester was being relocated to a 600-acre site in the nearby town of Gates, NY. The site, known as the Elmgrove Plant, became the center of Kodak's U.S. equipment manufacturing until its ultimate sale 32 years later, in 2000. The previous equipment organization, the Kodak Apparatus and Optical Division was renamed the Kodak Apparatus Division (KAD).

Since it was a common feeling among equipment folk at KAD back in the 1960s that we were "second class" citizens compared to the chemists and film people at Kodak Park (on Lake Avenue), we were rather pleased with this development, as it signaled that maybe we were beginning to get some respect, relative to our counterparts at "the Park." It wasn't long before Elmgrove was my new base of operations.

1969

Kodak Colorado Plant

Construction began on Kodak Colorado Division—a manufacturing unit for films and papers, located in Windsor, Colorado. Later on, our thermal print media was made there.

Apollo Stereo Camera

A very special stereo camera made by Kodak accompanied *Apollo* astronauts Buzz Aldrin and Neil Armstrong when they set foot on the moon. This camera produced three-dimensional close-up still pictures of the lunar surface for later scientific study.

End of the Sixties

Kodak ended the 1960s proudly and solidly in the picture business; nobody I knew thought otherwise. If you had asked Kodak employees about it back then, they would have told you that Kodak's business was pictures; they might have said "memories," or "images" but, even then, they really meant pictures. Revenue exceeded $2 billion, and the number of worldwide employees went over 100,000.

3

THE SEVENTIES

70s

I completed my coursework and, after going down a couple of research blind alleys, started working on a thesis for my doctorate in electrical engineering. Since my two-year award program was over, I returned to work at the KAD Research Lab, led by Dr. Bill Haynie. Dr. Gary Conners was there too, and we worked on a number of interesting development projects including liquid crystals and polishing aspheric optics. Then I became interested in electrophotography, the complex, multi-step process of putting toner on paper in an image-wise fashion that was the basis for "xerography" (or dry writing), as the Xerox Corporation called it. It was getting a lot of attention at Kodak, mostly in secret, and looked like a great way to make pretty good, low-cost images. I started making a mathematical image chain model for high-quality solid area development, and became quite involved in the technology, working in developing electrophotographic subsystems throughout the 1970s, leading to the *Ektaprint* copier line launched in 1975. On lunch hours, nights and weekends, I worked on my thesis, *A Geometric Optics Study of Rays in Inhomogeneous Guiding Media*, and completed the Ph.D. in 1971.

1972

Pocket Instamatic Camera

Kodak shrank the popular *Instamatic* camera to pocket size with the introduction in 1972 of five different *Kodak Pocket Instamatic Cameras*, using a new, smaller, *Kodak 110 Film Cartridge*. The line was so compact and popular with consumers that more than 25 million cameras were produced in less than three years. This was another dominant design. At this point in time, the terms *Instamatic* and *camera* were almost interchangeable to the average person.

Fig. 1972.1 *Pocket 20 Instamatic Camera (GEH)*

The camera featured a little *flashcube* that rotated around to provide four flashes in one package for customer convenience (people had sometimes burned their fingers on the flash bulbs when replacing them one at a time in the previous *Instamatics*).

The *110 Cartridge* was significantly smaller than the previous *126 Cartridge*. Here they are, side by side for comparison:

Fig. 1972.2 *126 (left) and 110 Instamatic Cartridges (GEH)*

This trend of ever smaller film frame sizes was continued a decade later with the *Kodak Disc Camera*. However, the judgment of history is that the *Disc* went a little too far, as you will learn in 1982.

1973

Spin Physics

In order to help develop tape-based digital storage technology, Kodak bought Spin Physics, a San Diego-based company started in 1968 that produced quality magnetic recording heads. The operation was eventually euphoniously renamed Kodak Research Labs/La Jolla/San Diego, but naturally, we always called it Spin Physics. Dr. James U. (Jim) Lemke was the principal, and when I met him several years later, we talked about ways to get Kodak to invest in and develop new electronic imaging technologies for future picture customers. Jim had me out to his all-teak house on the beach in Del Mar to discuss all the future electronic products we could conjure up together. He got a fundamental patent on the "micro-gap recording head," an approach for high-density recording that was incorporated into the many subsequent video cameras and video recorder/players that used magnetic tape.

Later, in 1980, another Spin Physics engineer, John Bagby, designed a novel high-speed electronic motion picture camera that was sold for use in slow-motion engineering analysis applications. Spin Physics was shut down about 25 years later, in 1998.

1974

Start of the Disc Camera Program

In 1973, Don Harvey, a great inventor and development engineer at KAD, began contemplating a pocket-sized camera that would have a small film size, available light capability, and motorized film advance. By inquiring into the rate of film emulsion research at KRL, he calculated that Kodak would come out with about one additional stop in speed/grain every ten years, and so he projected that he could have a new format about 2/3 the linear size of the *110* model by the early 1980s. (A film cartridge like the *110* was quickly ruled out as being too bulky and awkward to handle.) The small negative size that he envisioned suggested a small round disc of film, which seemed promising, and along with his assistant Nestor Rodriguez, Don unofficially started the "Dial Camera Program." The two men decided to use a plastic hub on the film disc for indexing and to provide a lock to avoid double exposures. The original idea was to have ten pictures per disc. They also contemplated a small mirror in a single lens reflex (SLR) configuration.

Dana Wolcott joined them in June 1974, and contributed numerous working models of both the camera and the film pack. Dana was responsible for what Don called the "handsome turtle-backed styling dummy" that "caught the fancy of many," meaning that it sold the program.

Fig. 1974.1 *Dana Wolcott's Style Model That Sold the Disc Program (DW)*

The Dial Program officially became P-157 in December 1974. (P-programs were Kodak's way to label significant programs and provide a high level of security—you couldn't know what P-157 was unless you were cleared). The program evolved into Kodak's *Disc Camera* project, and became a huge undertaking ... more about it in 1982 when it was launched.

1975
Ektaprint Copier-Duplicator
When Kodak introduced the *Ektaprint 100 Copier-Duplicator* in 1975, it received immediate industry acclaim for its high-quality copies and user convenience. It was one of the first products anywhere to have an on-board microprocessor, the *Intel 8008*.

This product caused a great stir at Xerox, since it employed an organic photoconductor belt, a recirculating feeder, and a conductive magnetic brush development process. All three of these innovations were being developed by Xerox at the time, but Xerox was slower in bringing out these new ideas, partly due to its excessive comfort level with the products it already had. Now the comfort was gone: the *Ektaprint*'s new organic photoconductor belt allowed users to get away from expensive selenium drums; its recirculating feeder provided completely finished sets without a sorter; and its conductive development process gave high-quality solid area tone reproduction. The

Fig. 1975.1 *Ektaprint 100 Copier–Duplicator (EKC)*

Ektaprint's image quality was clearly superior to the existing Xerox copiers that only reproduced the edges of solid areas on original documents.

The basic *100 Model* (above) had a platen, but other models had the recirculating feeder. The recirculating feeder innovation, designed by Matt Russel and Ron Holzhauser who worked in my Subsystem Equipment Development group in Copy Products, was a great feature in the days before the types of electronic scanner/copiers we have today. Here's a front view:

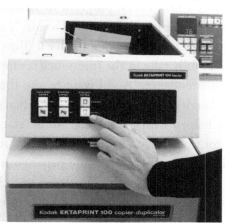

Fig. 1975.2 *Front View of the Recirculating Feeder (U of R)*

The design was clever in its simplicity, employing an oscillating vacuum roller to grab the bottom sheet in a stack of originals without ever risking a double feed. This type of vacuum roller is common today in paper handling systems. Matt Russel's invention got him elected Inventor of the Year by the Rochester Patent Law Association in 1988, and put him in the Copier Hall of Fame in 1992. If you could look inside, you could better see how it worked, but since you can't, it is diagrammed below:

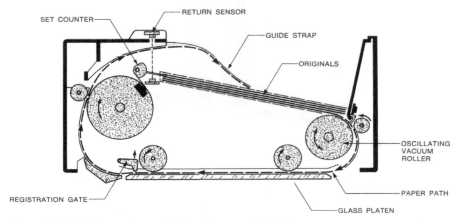

Fig. 1975.3 *Recirculating Feeder Diagram (U of R)*

Later in my career, after leaving Kodak, I was having lunch with a group of scientists on the Academic Advisory Board at RIT, trying to help define the scope of RIT's first Ph.D. program in Imaging Science. Mark Myers, then head of Xerox Research Labs, said to me, "I want to thank you for coming out with the *Ektaprint* in 1975; it gave us the shot in the arm we needed." Before I could fully take in the (presumed) compliment, he added, to the delight of all in attendance, "And I want to also thank you for letting us catch up!" I had to laugh at his biting wit also, but then said that although I was pleased to have helped with the first part of his statement, I was definitely not a contributor to the second turn of events. More laughs ensued.

In fact, in the late 1970s, I wrote a white paper entitled, "The Promise of *Ektaprint* in the Eighties," that was considered so radical that my boss buried my one and only copy. Ideas included scanner capture, digital storage, laser exposure, color, more compact copiers, fax capability, etc. Wish I had a copy today.

My friend and colleague Tony Ateya, who worked with me in subsystem equipment development in Copy Products, reminded me about an innovative follow-on product to the *Ektaprint* line. This was a project headed up by Bill Cavagnaro to produce a single-pass duplex copier, a big step forward from the double-pass methods available at the time. In fact, Bill got a patent on the paper turnover scheme that made it possible, and Tony got a patent on a clever air transport system that could move a sheet of paper with unfused toner on both sides into a set of fuser rollers without any mechanical objects touching the sheet and possibly degrading the image. The *Kodak Ektaprint 250 Copier-Duplicator*, introduced in 1982, provided a substantial advance in the technology of single-pass duplex copying.

Kodak was always somewhat ambivalent about the copier business. One reason was that in the copier business, you invest money to build copiers, and then get your money back by gradually charging "by the click," or copy. This is a different business model than selling film to a drugstore, say, where once your film factory is in place anyway, you get your money back sooner, after simply delivering the product. Another reason that Kodak was ambivalent about the copier business was because the image quality of the electrophotographic process was always considered inferior to film, although surely good enough for the intended application of copying documents. And, of course, copiers don't make as much money as film. Nor did much of anything else.

Electronic Still Camera Prototype

Back in the KAD Research Lab, Gareth Lloyd and Steve Sasson invented the first electronic still camera prototype. Gareth was a personal friend of mine. Although I was absorbed with electrophotography at the time, he and I were working in the same lab, and once during a lunch conversation we were talking about a new electronic image sensor, a Fairchild 100x100-pixel, monochrome charge-coupled device (CCD), and wondered what kind of electronic images it could make—after all, it had 10,000 pixels; 100 up and down, and 100 sideways! (A CCD is a type of electronic "eye" that was used for all early electronic cameras, and the term "pixel" is slang for a picture element. In this case "pixel" refers to a single detector in a CCD containing the image data for a pixel in the resultant picture.)

In 1974, Gareth creatively challenged Steve, a newly-hired Kodak engineer, fresh from RPI, to somehow make a prototype camera with this CCD that would capture an image and store the image electronically (certainly none of us knew exactly how at that time). Steve worked diligently on this project and actually got it to work in 1975, ably assisted by a superb technician, Jim Schueckler. His prototype was about the size of a large toaster, weighed almost eight pounds and recorded an image on an audiocassette tape in about twenty seconds.

Fig. 1975.4
Steve Sasson's Prototype Electronic Still Camera (EKC)

It was mostly a curiosity at first, but Gareth and Steve filed for a patent in 1977 called, aptly enough, ELECTRONIC STILL CAMERA, and the patent was issued in 1978. The prototype languished for years, more or less forgotten. Steve was not encouraged to speak about it except at a few early management demos, but luckily, Steve decided to bring it home to stash in his basement when it was about to be tossed out in a lab move. When Kodak patents began to get increased scrutiny around 2001, Steve was officially asked to speak about it. Now it is considered a one-of-a-kind museum piece, and a proud example of Kodak's innovations in electronics development, which have certainly benefitted the world.

This is as good a place as any to explain the difference between an electronic camera and a digital camera, as a lot of people are confused about it. For starters, electronic is a broader term than digital. Digital is, in fact, a subset of electronic. An electronic camera would not use film to capture the image, but rather an electronic sensor, like a CCD. However, a CCD is not inherently a digital device, because it spatially samples packets of photons at each pixel and converts them to electrons that are output as an analog voltage. If the camera were to store that voltage signal as an analog video signal, then the signals would stay analog, and it would just be called an electronic camera. In order to store that same image data in a digital format, those analog signals would first have to be converted to digital signals, using an analog-to-digital (A-to-D) converter to create the needed step-wise discrete signals from the smooth analog signals. A camera outfitted with such an A-to-D converter and capable of storing the resultant digital data is called a digital camera. So all digital cameras are electronic, but not all electronic cameras are digital. In these early days, before digital chips became readily available, any camera that didn't use film was referred to as "electronic."

Steve Sasson was trying to make his camera portable, so rather than use bulky analog circuits and video recording apparatus, he wanted to "freeze time" by quickly converting the analog signals to digital signals and storing the output (four bits per pixel) in twelve 4096-bit dynamic memory chips which had just become available, and then taking his time (23 seconds) to reliably read the resultant digital data onto a magnetic tape cassette. Steve needed a reliable storage means on tape because he knew that his prototype camera would be compared to cameras that enjoyed the reliability of film to store images.

So even though the patent was titled ELECTRONIC STILL CAMERA, because that was what we called all the early "non-film" cameras, this camera was actually a digital camera also. The decision to create and store four digital bits per pixel on a cassette tape was a brilliant stroke by Steve. At the time, most young engineers given the same assignment (to make a camera that didn't use film but instead used a CCD to capture images) would have simply stuck with analog video circuits for image storage. But Steve was intrigued with all the new digital electronics becoming available, like memory chips and A-to-D converters, and decided to take a chance and store the images digitally. He personally received the National Medal of Technology and Innovation on November 17, 2009 from President Obama for "the invention of the digital camera, which has revolutionized the way images are

70s

captured, stored and shared, thereby creating new opportunities for commerce, for education and for improved worldwide communication."[3] Not too shabby.

Steve also built a separate playback unit in a bulky but flexible microprocessor development system from Motorola, but the playback system no longer exists because the development system, being considered expensive at the time, was reused for other projects. Here it is displaying one of our favorite test images of the time, the "Boy & Dog" picture.

Fig. 1975.5 *Steve Sasson's Playback Unit (EKC)*

The playback image of the Boy & Dog got criticized a lot back then, as were all video images, especially when directly compared to the original photographic target, shown at the top of the next page.

You see, Kodak people were used to having to look very closely at image reproductions to tell the difference between the original and the reproduction. Unflattering comparisons between early electronic images and *Ektachrome* were common, although ultimately shortsighted. One of the toxic side effects of this sort of comparison was that it led our management to firmly believe that electronics would *never* catch film for quality.

I can tell you we were all rather impressed with this prototype camera at the time, and I expect some of my readers will find that hard to believe, because electronic digital cameras are everywhere today, and taken for granted. For example, people often ask Steve if they can see a print of the Boy & Dog picture, and Steve has to tell them "No," because there were no electronic printers around in 1975.

Fig. 1975.6 *Boy & Dog Photo Target and TV Display (EKC)*

An interesting artifact (below) will to help you appreciate how rudimentary electronic imaging technology was back in the 1970s. It was made by Ed Granger, another imaging whiz in the KAD Research Lab.

Around the time of Steve's invention, we were working with a Calcomp pen plotter to make charts for reports. Ed took a print of the famous Boy & Dog photograph, scanned it on a rather fancy and expensive scanner at KRL called the K.S. Paul scanner, made an electronic signal, and wrote some Fortran code to produce the digital halftone image below from that signal on the pen plotter. Ed's clever halftone used the presence or absence of four

Fig. 1975.7 *Ed Granger's Boy & Dog Print (EG)*

lines per pixel horizontally, and the resolution limit of the plotter to draw a short line vertically to approximate the picture density. I believe that was the first time I saw an electronic print of any sort, and we all thought it was great! Doesn't look too great today, of course.

So the first digital still camera had understandably low quality by today's standards as it had only a 0.01 megapixel CCD; nevertheless, it was the first digital camera actually built, and it worked! In Steve's KAD Research Lab report on the project, which I claim still reads well today, there is an interesting (under-) statement:

> *The camera described in this report represents a first attempt demonstrating a photographic system which may, with improvements in technology, substantially impact the way pictures will be taken in the future.*[4]

Steve went on to have a great career at Kodak, and our paths were to cross again in other electronic developments in the next decade. Our paths crossed after we retired also, as when Steve gave an excellent talk on his invention at the U of R Laser Lab in 2010. This picture was taken 35 years after the prototype was created. The prototype looks the same; we don't.

Fig. 1975.8 *Steve and Brad at the U of R Laser Lab in 2010 (SS)*

Bayer Color Filter Array

Kodak researcher Bryce Bayer invented a filter array that became the *Bayer Pattern Color Filter Array*. CCDs were the "eyes" of every early electronic camera, and in 1975 they were all monochrome. Bayer's filter array allowed

CCDs to capture color picture data. An interpolation scheme was then used to reconstruct the full-color image. The fundamental order in which dyes are placed on the image sensor with this array (green-red/green-blue) is still in use today in most electronic cameras and camera phones, although many variations have been devised. The reason that this idea is so clever and practical has to do with the response of the human visual system to color, which peaks in the green; so, in this pattern array there are two green pixels, one paired with red and the other one with blue. I first met Bryce in a Probability Theory class at the U of R, and again met him at the Kodak Research Labs some years later; in both contexts it was apparent that he was a really sharp guy.

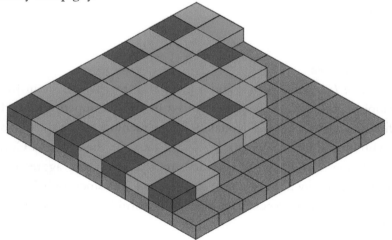

Fig. 1975.9 *Drawing of the Bayer Color Filter Array (Colin Burnett/Wiki)*

Bayer's color filter array is another great example of how Kodak electronics technology changed the world for the better.

Supermatic Film Video Player

In 1975, Kodak also developed a *Supermatic Film Video Player* that read *Super-8* film and produced a standard NTSC (National Television System Committee) video signal for display on television, either for a single TV or a closed-circuit network. This product was a low-cost telecine system, that is, one that converts film images into video.

This hybrid product (having both film and electronic components) competed for a while with the newly-emerging videotape recorders, which didn't use film at all but recorded analog signals on the tape. The *Film Video Player* used a flying-spot scanner (like on the *Lunar Orbiter*) to scan the film, and produced a TV signal of 60 fields per second from the 24 frames per second nominal film rate. It did this with a clever "two-three" scheme, where one frame is scanned twice while the next film frame is scanned three times, providing a 5/2 increase in frame rate (over the original 24 frames per second), or, 60 fields per second.

The *Film Video Player* was an easy way to get low-cost *Super-8* mm film movie images into a video format. You no doubt have seen many TV programs that used this telecine innovation (although you may have never realized it).

1976
Instant Camera

New *Kodak Instant Cameras*, and a print film to go with them for self-developing color prints, were announced in 1976, competing directly, for the first time, with Polaroid. Kodak had been manufacturing the first black-and-white instant film for years at Kodak Park for the Polaroid Corporation; but now Kodak felt it was time to compete more aggressively. In hindsight, that may not have been such a good idea since, unbeknownst to anyone at Kodak, a successful lawsuit by Polaroid was twelve years away (1988).

Fig. 1976.1 *EK6 Instant Camera (GEH)*

Sony Betamax VCR

The Sony *Betamax SL-7200* videocassette recorder (VCR) was announced using the now obsolete "Beta" magnetic tape cartridge. This was the first VCR introduced in the U.S., and was a big deal in the market, selling at $1,295. However, the first Beta tape allowed only one recording speed, and the capacity of each tape was just one hour—OK for TV recording, but movies tend to last longer than an hour. Sony corrected this the following year with the *SL-8200* model, which could handle a two-hour recording; however, the competition, led by JVC (a Japanese consumer and professional electronics company) had a larger, "VHS" cassette, available in both two- and four-hour recording times. By 1980, the VHS format had captured 70% of the U.S. market, and the resulting VCR war is still a classic marketing business case.

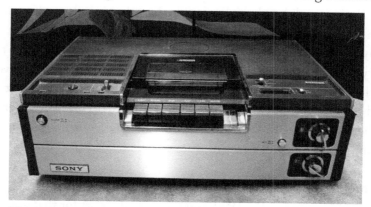

Fig. 1976.2 *Sony Betamax XL-7200 (GUSA)*

Kodak's Secret VCR Project

It turns out that before the *Betamax* was announced, Kodak had already been secretly developing a VCR using magnetic tape, called by the code name "VP-501." The idea was that since Kodak Pathé in France was already making high-quality magnetic audiotape, and since Spin Physics in San Diego could make high-quality magnetic heads to read tape, Kodak should be able to produce a high-quality VCR. Key people on this project team were Nevil Davy, Ed Granger, Fred Metildi, and a professor at RIT named Harvey Rhody, whom I met after I retired. The product never made it to the market though, partly because the video quality wasn't impressive back then as I have discussed, and partly because a Kodak executive at the time opined that nobody would pay a thousand dollars to watch Howdy Doody on TV. Turns out he was wrong, and, of course, the price of VCRs dropped with increased manufacturing volumes.

1978
Eastman KodaPak Thermoplastic Polyester
In 1978, the Eastman Chemical Division, located in Kingsport, Tennessee, introduced *Eastman KodaPak Thermoplastic Polyester* for use primarily in manufacturing beverage bottles. I had met only a few folks from what we then called "Tennessee Eastman" at occasional meetings attempting to look for "synergies" between the Kingsport employees and those in Rochester. I never personally thought that those meetings were very successful. The plant in Tennessee did make polyester film base, but its sales to Kodak were less than 10% of its worldwide sales.

Anyway, Kodak was now in the "pop bottle" business! I really didn't stew on it very much at the time, being rather obsessed with electronics development. However, in retrospect, I wonder what other projects might have emerged if Eastman Chemicals had been sold earlier (that is, earlier than 1994 when it was finally spun off), and the money from the sale invested instead in the "picture" business?

End of the Seventies
Kodak's Centennial Year was 1979, and the annual report for that year made many proud references to the previous 100 years. Sales had steadily climbed during the decade, and net earnings hit just over a billion dollars for the first time in the company's history. Worldwide employment was 126,300. It was the good times.

The only electronic imaging products mentioned in the annual report were the improved *Ektaprint Copier-Duplicators*. The brief comments on research were organized into two categories: silver halide research and non-silver research. The latter mentioned photoconductive films, photoresists for microelectronic fabrication, and energy-responsive polymers for optical disc materials. I have always disliked definitions of something as non-whatever, not just for the obvious logical problems, but because it often has a polarizing tone (e.g., "non-alcoholic.")

Not much doubt about it, at the end of the decade, Kodak was still in the picture business with film as the mainstay source of profits, but with new technology emerging, like magnetic tape, CCDs, optical discs and microprocessors, changes in the imaging business were underway.

Fig. 1978.1 *Eastman KodaPak Thermoplastic Polyester (EKC)*

4

THE EIGHTIES

1980
Ektachem Analyzer

As the decade started, the company entered the clinical diagnostic market with the *Kodak Ektachem 400 Analyzer*, a method of dry-chemistry blood serum analysis. My good friend Gareth Lloyd (recall the first electronic camera in 1975) was working on the electronics of this hybrid chemical/electronic product during development, and gave me a peek at it once when I told him I was curious. The product, born in KRL, was ingenious. A drop of blood serum to be tested was placed on a special chemically-coated "slide"; the resultant optical color variations indicated the amount of, say, glucose that was in your blood. The saving grace, of course, was that the slides were consumables. This entire product line was eventually sold to Johnson & Johnson in 1994.

Fig. 1980.1 *Ektachem 400 Analyzer (EKC)*

Spin Physics Motion Analysis System

Motion analysis of fast-moving objects requires capturing many images at high speed, and then reviewing these images slowly to see what was really going on, beyond the capability of human vision. Applications range from sports to safety to engineering: an athlete throwing a baseball, an automobile tire hitting a pothole, rapid fluctuations in high-speed machinery. Kodak's *SP-2000 Motion Analysis System,* designed by John Bagby at Spin Physics and introduced in 1980, used a very fast CCD, high-speed electronics, and advanced tape-recording technology to perform high-speed video recording at rates up to 2,000 frames per second. This enabled powerful slow-motion analysis for both industrial and scientific applications.

Fig. 1980.2
Spin Physics SP-2000 Motion Analysis System (EKC)

1981

IBM PC Model 5150

On August 12, 1981, IBM introduced its *PC Model 5150*. It had an Intel 8088 processor running at 4.77 MHz, two floppy disk drives at 160 KB each, and a PC-DOS (personal computer-disk operating system) boot disk from the as yet unknown Microsoft Co. And it had no hard drive! The computer had 16 KB of random-access memory (RAM), at a base price of $1,565. All of us geeks were clamoring for one of these, and soon Kodak became a "Big Blue" house, in that, if a computer wasn't from IBM, you couldn't buy it. During the first four months, IBM sold 35,000 of these computers, and 800,000 by the end of 1982.

Fig. 1981.1 *IBM PC Model 5150 (Boffy b/Wiki)*

Sony MAVICA

But the big shocker, as far as Kodak was concerned, was Sony's announcement in 1981 of the *Sony MAVICA* (MAgnetic VIdeo CAmera). It resembled an ordinary 35 mm camera, but was actually a color electronic camera with a 570x490-pixel CCD, capable of storing 25 frame images or 50 field images on a 2-inch floppy disk called a *Mavipak*. (An NTSC video frame consists of two interlaced video fields). It was not a digital camera because it stored

Fig. 1981.2 *Sony MAVICA (Morio/Wiki)*

analog video signals on the still video floppy disk. However, since the video floppy looked like a shrunken floppy computer disk (which itself was digital), many people became confused and thought the *MAVICA* was a digital camera. Initial pricing was $660 for the camera and $2.60 for a disk. Sony Chairman Akio Morita prophetically told a Tokyo news conference that the new camera "will make conventional chemical photography and development obsolete."[5]

Morita-san was a very sharp guy, whom I met once briefly later on. The actual sale of cameras in Japan was 18 to 24 months away, but this announcement caused quite a stir at Kodak, initiating a chain of events that was to change my life, ready or not.

1982

Disc Photography

Kodak launched *Disc* photography with a line of high-quality, compact, "decision-free" cameras built around a rotating disc of film using the new *T-Grain* Kodacolor film technology. Unfortunately, although the film was excellent, it was not quite up to the expected task at that time, considering the very small 8x10½ mm film format. The *Disc* film format area was slightly less than 10% of the 35 mm format area, so the pictures often appeared somewhat grainy compared to the increasingly popular 35 mm cameras. Nevertheless, it was an innovative, easy-to-use snapshot camera system that met many customer needs, and was commonly perceived to about equal the previous *110 Instamatic* format. There were complaints inside Kodak about the grain problem, but by then the project inertia was overwhelming and it went charging ahead. I was appointed coordinator for *Disc* camera manufacturing, and got deeply caught up in the project frenzy.

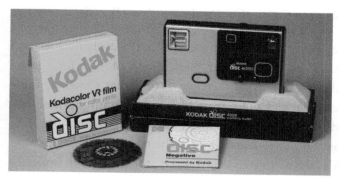

Fig. 1982.1
Disc 4000 Camera
& Disc Film (GEH)

What became the *Disc* program, you may recall, was started at KAD by Don Harvey, Nestor Rodriguez and Dana Wolcott back in 1974, originally with ten frames per *Disc* and a binocular camera format. Technology marches on, incessantly, and usually takes a little time. Some of the other Kodak innovations in the *Disc* camera system besides the film and *Disc* geometry were the 12.5 mm focal length f/2.8 glass optics, the lithium battery, and integrated circuits for automatic exposure control. Each of these three innovations was a first, and deserves some additional comment.

The Kodak lens was actually revolutionary, as it employed aspheric optics molded from glass for the first time in a consumer camera. Aspheric glass optics were known to be superior to employ in lens designs, but until the *Disc* camera, glass aspheres were only used in commercial applications, due to their cost (previous consumer aspheres were molded from plastic, which is inferior to glass). The *Disc* lens was all glass, consisting of an airspaced triplet with a field-flattening fourth element, as seen in the cross-section below:

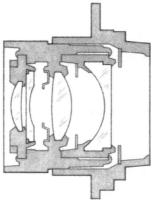

Fig. 1982.2 *Cross-Section of the Disc Camera Lens (Pop Pho)*

The third surface from the front (on the left) is the aspheric surface, although you can't really see it. The short focal length of 12.5 mm gave greater depth of field than either the *126-* or *110-*format cameras. This meant a fixed-focus lens that was sharp from four feet to infinity. The short focal length also made possible a larger aperture of f/2.8, with about 2½ stops more light-gathering power than the *110*. The resolution was close to the best possible with a lens of a given aperture, i.e., the diffraction limit.

The batteries were another first, being a pair of three-volt high-power lithium batteries with a five-year life expectancy, far more than the predicted usage by consumers. If the batteries wore out, a dealer would replace the whole camera, i.e., a lifetime guarantee. This technology is now common in button- or coin-sized cells, but the *Disc* was the first camera to use it.

The *Disc* camera's automatic exposure control capability was also a first. One integrated circuit in the *Disc* camera detected the amount of light on the scene with a threshold of 125 foot-lamberts. Above that, the camera shot at 1/200th of a second at f/6; below that, it shot at 1/100th of a second at the full f/2.8. Below the threshold, the electronic flash always fired. A second logic chip controlled the aperture, the flash charging and firing, and automatic film advance. Dana Wolcott told me later that a more continuous exposure control would have reduced the grain problem, but the project decision was made to forge ahead.

Eventually, Kodak introduced an improved *T-Grain* film, but it was not used in the later *Disc* cameras. I inquired about this, and was told that we had lost the marketing momentum. I guess that was true, especially compared to the growing 35 mm camera popularity. However, while I'm no marketing expert, I've always felt that if you have a better product feature or capability for an existing product line, you should implement it as soon as possible for the customer's benefit. I was rather sad about this one, for I knew and respected the *Disc* camera's inventor, Don Harvey, and the cameras that were produced were slick, near-diffraction-limited little jewels. I believed that using the new improved film in these lovely cameras was a no-brainer. Guess it wasn't, somehow, but I still feel I was right on that one.

At the 1982 Photokina show in Germany, the Kodak Research Lab people demonstrated a technology prototype that would allow pictures on *Disc* film negatives to be displayed on a television screen. It was cool, and got some press, but never came to market. *Disc* cameras went out of production in January 1988, after only six years.

1983

Mead Digital/Diconix/Scitex

Mead Digital Systems, founded in 1972, introduced the *Dijit*, its first commercially available inkjet product, in 1973. A high-speed, continuous inkjet printer, the *Dijit* could print variable data, e.g., "Dear Mr. Constantinides, you may have already won…" Kodak acquired Mead Digital in 1983 and renamed it Diconix. It was transferred to Kodak's Copy Products division in 1988, and given the uninspired name of Dayton Operations.

Soon after, when I became general manager and vice president of the Printer Products Division, Dayton Operations reported to me, and we used the Diconix name for products. Dayton Operations was developing high-speed, higher resolution and process-color continuous inkjet technology, which I encouraged and supported because it was clearly the future of variable data digital printing. In the long term, I believed, it was the future of printing, period. I still do.

A year after I left the company (1993), Kodak sold this division to Scitex—I was no longer there to protect the Diconix people while they developed their incredible technology. I was not pleased, and knew this was a poor decision, but unfortunately I was a lame "Koda-duck" at that point. You may find it hard to believe what happened to Scitex in 2004.…

Consumer Products Development

In 1983, I was running the Consumer Products Development group at Elmgrove Plant, working on electronic imaging products that could be commercialized, partly as a response to the *MAVICA*. These involved both still video images using small floppy magnetic disks, and motion video images using compact 8 mm magnetic tape. In order to determine whether some suitable Japanese companies could partner with us, I went on a two-week trip led by Dave Greenlaw of Corporate Commercial Affairs to visit Sony, Hitachi and Matsushita. We selected Matsushita for the 8 mm video project but were very impressed with the electronics work going on at Sony and Hitachi as well.

1984

New Business Units

This was the year Colby Chandler shook things up at Kodak by forming 17 separate business units, in order to get the company to move more quickly in this more competitive time that was clearly upon us. One of these units was the Consumer Electronics Division (CED), led by an up-and-comer named

Dan Carp. We in Consumer Products Development all suddenly became part of CED, and I became Dan's development engineering guy.

Diconix Portable Inkjet Printer

In 1984, Kodak introduced the *Diconix 150 Plus*, the world's first portable inkjet printer, using drop-on-demand (DOD) technology. It was only 2x6½x10¾ inches, and ran for 12 hours on 5 C-cells, cleverly placed inside the platen roller to save space. It was priced at $499.

Fig. 1984.1 *Diconix 150 Plus (KBP)*

Kodavision Camcorder

Kodak/CED entered the video market with the *Kodavision Series 2000* 8 mm video system in 1984, and introduced Kodak videotape cassettes in 8 mm as well as Beta and VHS formats. The *Model 2400* had autofocus.

Fig. 1984.2 *Kodavision Camcorder (EKC)*

The innovative design involved a "cradle" in which the camcorder was placed for playback of home movies and recording of television programs using the tuner/timer.

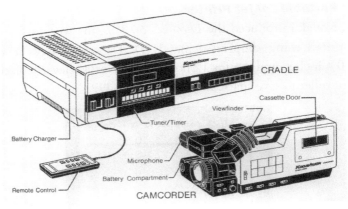

Fig. 1984.3 *Kodavision Camcorder and Cradle (KBP)*

Here is the camcorder placed in the cradle:

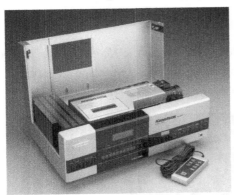

Fig. 1984.4 *Kodavision Camcorder in its Cradle (EKC)*

This was the world's first 8 mm camcorder, designed in cooperation with Matsushita in Japan and built in Japan. The Japanese were ahead of Kodak in the production of helical-scan magnetic recording systems, and the exchange rate was over 200 yen per dollar, so it was considered difficult to make money manufacturing these products in the U.S. Nonetheless, I felt it was a good start. In ten years, the exchange rate would drop to about half, or a little over 100 yen per dollar.

Apple Macintosh Computer

Another monumental electronic milestone in 1984 was the introduction of the *Apple Macintosh* computer, which had a friendly graphical user interface (GUI), icons, and a mouse, all radical innovations, originally stemming from Xerox PARC (Palo Alto Research Center). Although the *Macintosh* had no hard drive, it had a 3½-inch, 400 KB floppy disk drive and 128 KB of random access memory or RAM. It was priced at $2,495, and Apple sold 70,000 units in three months. Even though it was only black-and-white, it was imaging-friendly, and a huge advance in personal computers. To this day, people who work with images on their computers prefer *Mac*s. I thought it was a sweet product, and moved toward the *Mac* and away from *IBM PC*s as soon as I could, never going back to this day.

Fig. 1984.5 *Apple Macintosh Computer (CHM)*

1985

A Hybrid Approach

Kodak's 1985 annual report had a couple of interesting quotes in it from two top executives whom I admired, both promoting the "hybrid" approach to Kodak's future picture business, that is, the use of both traditional and electronic imaging in systems for customer benefit. J. Philip Samper, executive vice president and general manager of the Photographic and Information Management Division, said, "There are excellent growth opportunities for our current products, as well as for those future offerings that combine the benefits of photographic and electronic imaging." Dr. Edwin P. Przybylowicz, senior vice president and director of research, said, "Since it was started at Kodak in 1912, research has been a cornerstone of the growth of our company, concentrating on imaging sciences and leading to innovative products based on silver halide, electrophotography, and electronics." Clearly, these leaders thought we were in the "picture" (aka "imaging") business, and supported the hybrid approach, which we all understood would help pave the way toward electronic imaging systems of the future, while still allowing us to enjoy some film profits.

KEEPS and KIMS

In 1985, the company introduced two new image management systems—the *Kodak Ektaprint Electronic Publishing System* (*KEEPS*), and the *Kodak Information Management System* (*KIMS*).

KEEPS was a professional electronic publishing system sold internationally by Kodak from 1987–1992. As a fully-integrated turnkey system, *KEEPS* consisted of publishing software from Interleaf, computer hardware from Sun Microsystems, customized front-end software developed by Kodak that ran on Unix Release 4, and Kodak's high-end *Ektaprint* printers, scanners, and copiers. *KEEPS* produced WYSIWYG (what you see is what you get) output at near-typeset quality, while also offering document management and workflow tools for collaborative environments.

KIMS was a document storage and retrieval system that enabled users with large, active databases to manipulate information stored on microfilm, magnetic or optical discs.

A primary objective of these systems was to sell high-end imaging peripherals to existing business customers. They used Sun (*KEEPS*) and Digital Equipment (*KIMS*) workstations. Both systems were shut down around 1990.

Color Video Imager

In 1985, Kodak also announced the *Color Video Imager* (which we internally called the "*Trimprinter*") that could capture a video frame image from a video signal and make an instant print on *Kodak Trimprint Instant Film*.

We were all excited about this product, the brainchild of Dana Wolcott, because it was fun to use, and the *Trimprint Instant Film* was readily avail-

Fig. 1985.1 *Color Video Imager aka "Trimprinter" (EKC)*

able, being the same film as was used in Kodak *Instant* cameras. We took a picture of Colby Chandler with a prototype electronic still camera during a demo and made him a print on the *Trimprinter*. Mr. Chandler called me several weeks later to tell me that, unfortunately, the *Trimprinter* was never going to make it to market, because Kodak had just lost the Polaroid lawsuit and was going out of the instant photography business. It was very kind of him to call me, but we were all disappointed in this turn of events.

However, sometimes even bad events result in something good, especially in the world of R&D. It happened that just a few days prior to Mr. Chandler's phone call, Carl Kohrt and Scott Brownstein from the Kodak Research Labs had come to my office at Elmgrove Plant and shown me some sample prints that Scott had made using new thermal dye-diffusion technology. I thought they were awesome electronic prints, and right away wanted to support the commercialization of this exciting new print technology.

One of the reasons I was excited about this technology was that, unlike electrophotography or inkjet, which approximate each pixel with many smaller dots, the thermal technology made essentially a perfect pixel by depositing the correct amount of cyan, magenta and yellow dyes over the entire area of each pixel. It accomplished this by having the dyes on a dye donor ribbon, and an array of pixel-sized heater elements drove the dyes onto an awaiting receiver sheet that became the final print. This is in contrast to electrophotography and inkjet printing, wherein each pixel is approximated by clumps of toner particles or ink droplets, respectively.

So, before long, the core ex-*Trimprinter* team went over to building 59 at Kodak Research Labs to start work with Scott on a 4x5-inch prototype thermal printer. I told them, "The basement of B59 is hot in the summer and cold in the winter, but you will love working with Scott so much that you won't care." I dropped by KRL each Friday afternoon on the way home from the Elmgrove Plant to see the group, and they made terrific progress, eventually leading to many commercial thermal print products, including the photo kiosks. This became another example of Kodak developing a deep core competency in an important technology.

Modular Video System

As a follow-on product to the *Kodavision Camcorder*, we introduced the *Modular Video System* (*MVS*), a complete video movie system, including a camcorder, a video recorder/player, and a tuner/timer.

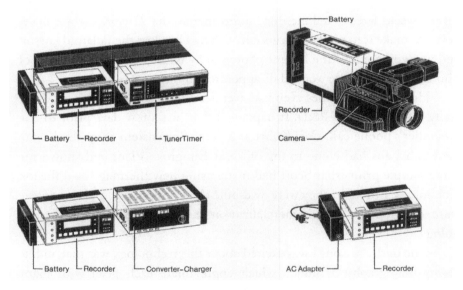

Fig. 1985.2 *Modular Video System (EKC)*

Eikonix Designmaster

Kodak acquired Eikonix in 1985 for about $56 million. The company made commercial imaging systems that scanned, edited and prepared images for printing. Its flagship product was the *Designmaster 8000*, a color electronic prepress system using a CCD in the scanner. This all-digital graphic arts system converted color transparencies and photographs into electronic high-resolution halftones for color printing. Kodak sold Eikonix in 1989, but kept the Eikonix name for some products.

Fig. 1985.3
Eikonix Designmaster 8000 (EKC)

1986

UltraLife Lithium Power Cells

In 1986, the company announced *Kodak UltraLife Lithium Power Cells*, the world's first 9-volt lithium cells for consumer use, and entered the general consumer battery market with a line of *Kodak SupraLife* batteries. The *Disc* cameras were the first products to use lithium batteries. It was considered virtuous that since film hung on hooks at drugstores, and cameras needed batteries, it was obvious to offer batteries hanging on hooks in drugstores also. (I am not making this up.)

World's First Megapixel CCD Sensor

During the same year, Kodak scientists invented the world's first megapixel CCD sensor, capable of recording 1.4 million pixels, sufficient to produce a 5x7-inch digital photo-quality print. The number 1.4 million is not as strange as it may seem, because we were developing thermal printing at the time, using a 200 pixel-per-inch design (and making great-looking prints), and in order to make a borderless 5x7-inch print without interpolation, you need about 5 x 200 x 7 x 200 = 1.4 million pixels.

As mentioned previously, the thermal printers made essentially a perfect pixel by depositing the right amount of cyan, magenta and yellow dyes on the receiver sheet over the exact area of each pixel, instead of just approximating each pixel with many dots of ink or toner, as with inkjet or electrophotography. (So, although it's confusing to many and pixels per inch

Fig. 1986.1 *The Kodak Research Lab's Path to a Megapixel CCD (EKC)*

are often mixed up with dots per inch, pixels and dots are not the same).

Even though we knew we were making photographic quality prints, because we could see them, the term "photo-quality" was not allowed by Kodak management to be used to describe electronic photo products from then up through the time I left in 1992, because it might encourage electronics to advance and hasten the demise of film, or, in other words, "It might hurt film." Also, there still was a strong belief in management circles that electronics could never catch film for quality, despite our many demonstrations to the contrary for various applications. I felt this was wrong, and an unhealthy attitude on the part of Kodak management, but had to deal with it. I recall we got by with "near-photographic quality" for a while, but Kodak couldn't control the world's progress.

Demise of Kodavision

Kodak quietly exited the *Kodavision* 8 mm camcorder business around 1985–1986. No reason was given publicly, but I know the message that got sent up to top management was that consumer electronics was a defective business, and should be abandoned. This was a message that the 19th floor of Kodak Tower (the highest floor, where the office of the CEO resided) wanted to hear, and all those who agreed with that message were smiled upon. Of course, the exchange rate was 200 Japanese yen per dollar back then, and electronic equipment profits never matched film, so I suppose it would have been extraordinary for an exec shooting for the top job to argue otherwise. I did not agree, and took a modest amount of abuse for my opinion. I felt, and stated strongly, that the talented and dedicated people at Sony, for example, did not think, as they got up each morning, that they were going off to feverishly work all day on a "defective" business.

One particular reason we in CED were not happy about this message was that we had on the drawing boards a compact 8 mm camcorder that was simple to use (I wanted one big orange "record" button) and could be held in one hand. The Sony 8 mm *HandyCam* was announced a year later, so perhaps we had a good idea. Of course, it wouldn't have made as much money as film, as I was constantly admonished. Sony still sells products called *HandyCams*, although of course they all use solid-state digital memory now instead of metal-particle 8 mm magnetic tape like in the 1980s.

Electronic Photography Division

Even though Kodak was down on "consumer electronics" at that point, we of the geek persuasion had argued for some time that commercial applications were really the right place to establish a foothold in electronic imaging. In 1985, Kodak reshaped its Consumer Electronics Division (CED) into a new business unit called the Electronic Photography Division (EPD), with me as vice president and general manager. Donna Malone, who was working for Dan Carp in CED, was my trusty secretary, and key early members of the technical team included Ed Brooks, Pete Lockner, Carl Schauffele, Dana Wolcott, Keith Hadley, Don Pophal, Andy Cooper, Bob Cosway, and many more. EPD, by design, concentrated on electronic products and systems that did not directly use Kodak's mainstay silver-halide photographic processes.

The systems we worked on could exchange images between traditional and electronic modes, and so were, in fact, hybrid products. We were rather fond of the "four-bubble" diagram shown below:

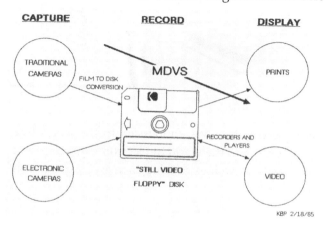

Fig. 1986.2

The "Four-Bubble" Diagram (KBP)

4 The Eighties

We started developing our own 3½-inch floppy disk for recording, but soon our focus was on commercial and industrial applications with a system of products built around the two-inch *Still Video Floppy Disk*, which was becoming an accepted standard in the world for still video photography. Our *Magnetic Disk Video System (MDVS)* could capture images with either traditional (film) cameras or electronic cameras, record the images on a *Still Video Floppy Disk*, and then display the images either as prints or on video screens. The *MDVS* name was a tad on the techie side (naturally), so, with some marketing help, we soon renamed it the *Kodak Still Video System (SVS)* as we began to approach the marketplace. Here is a hand-made color chart of *SVS* as we envisioned it in 1986:

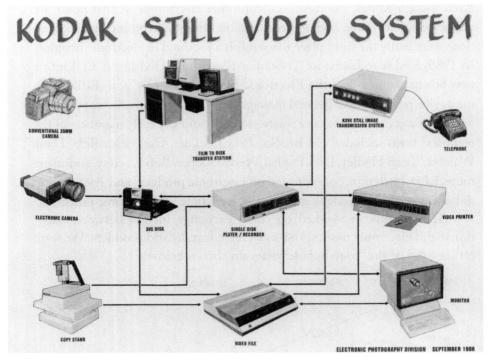

Fig. 1986.3 *Kodak Still Video System (KBP)*

In June 1987, this complete system was announced to the world (see 1987 below). The idea was to jump into the world of electronic imaging, and learn more about customer needs, as well as develop increased expertise in design and manufacturing. So we did.

Now that I was running one of the Kodak business units, I discovered an interesting toxic side effect of the new company structure: namely, that the units competed with each other rather vigorously. Pursuing R&D work related to electronic imaging technology that would benefit several business units became a daunting task, because each business unit wanted another business unit to pay for the R&D. Plainly speaking, less R&D expense (short term, anyway) meant greater net income and more executive bonus. I learned an important management lesson the hard way: companies have to be careful about providing incentives, because they might get exactly what they seek, and in the process, suffer some adverse consequences.

In late 1985, Tom Nutting and Keith Surdyke had developed a prototype electronic camera, a rather ungainly thing with a large over-the-shoulder pack containing our own 3-inch magnetic disk recording unit, but it worked. I took it to the Kodak main office sometime in early 1986, unannounced, to take pictures of Kodak executives as high up in the tower as I could get without being ejected. In front of an elevator, I met some chaps from Consumer Marketing who saw me and said, "Here comes Dr. Doom." It seemed funny at the time. Later on, when one of them called me on the phone, I picked up and answered, slowly and in my deepest and scariest tone, "DOOM here." I had to keep my sense of humor at this point.

Still Video Camera Design

During 1985–1987, Tom Nutting and his team continued developing a still video camera as part of the emerging *Still Video System* (*SVS*), attempting to use a prototype Kodak E3 interline transfer CCD with a video graphics array (VGA) of 640×480 pixels. In the early development stages, the electronic signal-to-noise (S/N) ratio was low due to dark current noise in the sensor. While the camera was coming together, including optics, exposure control, color balance, shutter, and the *Still Video Floppy* (*SVF*) magnetic disk recording system, the sensor S/N ratio was improving but remained a concern.

There were video (motion) cameras available that could be used to get electronic images into the *SVS*, and so a camera that recorded directly onto a *SVF* disk was not actually a critical component of the *SVS*. Kodak managers always showed a lot of interest in cameras due to their experience with cameras as devices made to burn film; however, they were frankly torn about aggressively promoting a Kodak-branded electronic camera because they thought it would hasten the demise of film. In the meantime, we designed and built about six models of what was called the *SV8300* for demo purposes and forged ahead, launching the other components of the *Still Video System* in 1987.

Fig. 1986.4 *The First Kodak Electronic Still Camera Design (GEH)*

This camera, shown above, never was produced, but it was the first Kodak electronic still camera designed for production. It was not a digital camera per se, because it recorded images on the *SVF* disk using analog video recording. Nevertheless, it was a neat binocular-style design we fancied at the time, and now, 11 years after Steve Sasson's prototype in 1975, it was a step along the path to improved cameras that "didn't use film." Although some very nice digital cameras were about to be created by Jim McGarvey for federal clients, and later for photojournalists, it would be almost another decade before Kodak finally introduced the "point-and-shoot" *DC40* digital camera for consumers, which also had a binocular design. It wasn't that we couldn't have moved faster; management was not comfortable with being aggressive about digital cameras.

As the Electronic Photography Division was getting started, Jack Flaherty, a Kodak Apparatus Division production manager who was instrumental in getting electronic manufacturing up and running at Elmgrove Plant, was retiring. At his retirement party, he "willed" me a very handsome portrait of George Eastman that hung in his office. I proudly hung it in my office at Elmgrove Plant, and one day saw a small note attached to the bottom of the picture frame. It read, "You're making a camera that doesn't use WHAT?" Visitors to my office got a kick out of it, and so did I, so I left it alone. Looking back, the joke perfectly reflected the times.

UltraTech

In 1986, Kodak gave one of its "venture" operations, Ultra Technologies, full status as a business unit in its Photographic Products Group. The company was created to develop Kodak's lithium nine-volt consumer-use battery (such as was used in the *Disc* camera). I wasn't keen on this turn of events. Ultra Technologies was sold in 2004.

Ektaprint Digital Printer

Around this time, laser printers were being challenged by solid state imaging alternatives, and at Graph Expo East 1986, Kodak demonstrated its *Ektaprint 1392 Digital Printer*, based on light-emitting diode (LED) technology. Instead of a laser with a polygonal mirror scanning back and forth across the xerographic photoreceptor, the *1392*'s long, thin array of tiny LEDs and gradient-index optics at 300 dots per inch created the page image, with fewer moving parts. While this Kodak product was meant for high-volume, high-speed applications, smaller scale LED products would soon begin to compete with lasers for digital printing applications in publishing. The *Kodak 1392* sold for $199,000, and could print 92 two-sided, 300 dots-per-inch (dpi) pages a minute.

Ektascan Medical Laser Printer

In 1969, Gary Starkweather at Xerox had invented what we today call a "laser printer." At that time, the laser was used to expose a charged photoconductor to produce an electrostatic image that subsequently attracted the charged toner particles. A rotating polygonal mirror accomplished the scanning motion of the laser beam. When electronic diagnostic imaging technology became available, Kodak marketing knew from experience that medical doctors would want to keep using their familiar 7 mil polyester-based film with a light box to look at diagnostic images such as x-rays, CT scans, etc. So, in 1986, Kodak introduced the *Ektascan*, a laser printer for medical images that exposed Kodak film from the original electronic medical images. It was a big business for Kodak for many years, and was another hybrid product. Carestream (Kodak's former Health Group) now sells the *DryView 5700 Laser Imager*, a descendant of the *Ektascan*.

Verbatim Optical Disc

Kodak bought Verbatim's plant in Limerick, Ireland in 1985 for about $175 million. The company made floppy magnetic disks, but armed with a £60 million Kodak investment, it became the first in the world to produce 14-inch optical discs. These discs were Verbatim's main storage business in 1986, with proprietary media, drives and jukeboxes (each holding up to 150 discs) for archiving. Capacity kept increasing: 10.4 GB in 1991, 14.8 GB in 1994, and 25 GB in 1996. Kodak got out of this business in 1998, after only three years.

Fig. 1986.5
A 14-inch Optical Disc (DM)

1987

Video ID System

On February 22, 1987, at the Photo Marketing Association (PMA)'s spring meeting and equipment show in Chicago, Kodak's Electronic Photography Division announced the *Kodak Video ID System* (*KVIDS*), at that time a new approach to making ID cards. It used a color video camera, an advanced

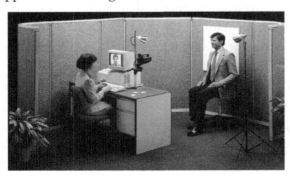

Fig. 1987.1
Our Kodak Video ID System at Work (EKC)

thermal dye-transfer printer, and digital image processing software to produce rugged, secure ID cards and badges.

The sample cards we gave out at the show looked like this:

Fig. 1987.2

Sample KVIDS ID Card (EKC)

These high-quality ID cards were very resistant to tampering or counterfeiting, and were produced in about 90 seconds. I was quoted in the press release as saying, "The microprocessor-controlled printer is capable of producing from blank printing material a fully-finished card, including logos and graphics, through the use of a programmed and easily interchangeable cartridge." That was the techie me, trying to help. The release went on: "If the image is not quite right for any reason, such as poor framing or the eyes being closed, it can be reshot,' noted Paxton. 'This means that expensive and time-consuming remakes can be greatly reduced.'"[6]

I took some grief over that (last) one: I was told by some Kodak folk that eliminating remakes was a dumb idea because it would reduce print sales (never mind that the customers might like it).

In October 1988, we showed the ID system to a very enthusiastic crowd at Photokina in Cologne, Germany. We made free custom luggage tags for attendees that were a big hit, such as the one below:

Fig. 1987.3

My Luggage Tag from Photokina 1988 (KBP)

That actual tag has been on my briefcase(s) from 1988 until today, and it's still in good shape! As for me, I'm 25 years older now, and a tad slower.

Still Video System

On June 18, 1987, at a press event in New York City, Kodak's Electronic Photography Division (EPD) entered the electronic still video market with seven products for recording, storing, manipulating, transmitting and printing electronic still video images for commercial markets. We also had the *Still Video Camera* as a demo model. (Here, "our" or "we" refers to EPD.)

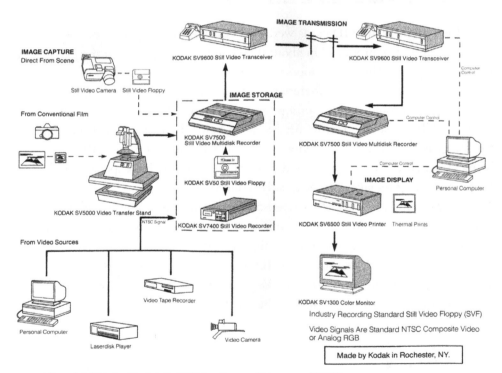

Fig. 1987.4 *Kodak Still Video System Diagram (EKC)*

Our recording medium was the *Still Video Floppy Disk*, which had fifty circular magnetic tracks on each disk for NTSC video; it held 50 field images or 25 higher-quality frame images, consisting of two interlaced fields each. On the next page you can see it with a familiar Kodak look.

Fig. 1987.5
Still Video Floppy Disk (DM)

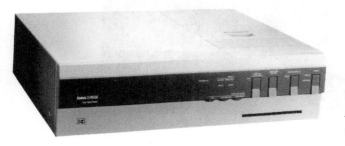

Fig. 1987.6
Kodak SV6500
Color Video Printer (SS)

Our printer, the *Kodak SV6500 Color Video Printer*, could produce an excellent color thermal print in about a minute from any NTSC video signal or computer output.

A good quality video signal was needed to make an acceptable print, and at that point in time the virtues of migrating from analog video to the digital recording of images were apparent. In fact, even as we were planning for the future, the *SV6500* as designed then could also print a digital signal from a computer, captured by a digital 512x512-pixel frame store. A more complete system solution for digital image storage would appear three years later as the *Kodak Photo CD* (1990). The *SV6500 Printer* also offered several special effects, utilizing the built-in *Centronics Port*: different aspect ratios, black-and-white prints from a color image, negative prints, a slick 4-up feature useful for multiple ID photos, and image edge enhancement.

We used the digital input to the printer to make dozens of a very nice sample 4x5-inch print that we were fond of showing around to anyone who cared because, frankly, it looked great. This print (see the next page) had a digital image area of 3.7x4.7 inches, and the thermal head resolution was 5.4 pixels/mm, so the total number of pixels used to make it was about 327 thousand, or a little less than a third of a million pixels.

This number has some real significance, actually. In 1987, some researchers were fond of telling management that there were the equivalent of 25 million "pixels" on a frame of 35 mm film, which gave credence to the

Fig. 1987.7 *Less than 1/3 million pixels! (KBP)*

thought that electronics would never catch up to film in terms of quality, which, of course, management wanted to hear.

In a briefing, and not without guile, I showed the print we had made to Colby Chandler and asked him how many pixels he thought were in it. I got a funny look from Phil Samper, but pressed on. Mr. Chandler was amazed when I told him it was (only) about a third of a million pixels. Further, I went on to say something to the effect that the 25 million pixel figure was kind of a "researchy" number, requiring perfect lenses, perfect focus, no dirt on the optics, ideal developing solutions at perfect temperature, and all the planets to be in alignment. I further clarified that ordinary customers who got their 4x5-inch prints from the drugstore never saw anywhere near that many pixels.

Mr. Chandler mentioned this to the head of research later, and I got in some more trouble for causing this regrettable "confusion" with upper management. However, I was trying to make the important point that the quality "bar" for electronics to catch up to the quality of film-based photography was not nearly as high as everyone thought. Steve Sasson told me recently that he felt it was the job of R&D to confuse management! Curiously, I immediately felt better.

We made other demonstrations along similar lines as the years went by, and it was always instructive, but in 1987 there was still the strong belief that, in terms of image quality, film would always be ahead of electronics. I was told, more than once, "It's nice Paxton, but it's not *Ektachrome*." Never mind that *Ektachrome* might not be the customer requirement! Then, as now, the subject of requirements and getting them right for the intended purpose is a profound topic, and the basis of good systems engineering, as well as smart business.

Another component of the *Still Video System* was the *SVS Player/Recorder*, designed by us to handle one disk at a time, to either record or play back video, and to be rather human-friendly.

Fig. 1987.8
SVS Player/Recorder (EKC)

It had some neat features that were easy to use, like a slide show mode and the option of adding frame/field numbers on the video image. We also had a multi-disk version of the player/recorder that held 30 disks in a cassette tray, so it held up to 1500 images; we called it our "electronic shoebox."

We also had an *SVS Transfer Station* for lab use, to scan traditional photos or film and record the images on a *Still Video Floppy,* helping complete the hybrid cycle.

Fig. 1987.9
SVS Transfer Station (EKC)

Some of our customer conversations indicated a strong desire to electronically transmit quality color images between two sites. This led us to develop Kodak's first all-digital product, the *SV9600 Still Video Transceiver*.

Fig. 1987.10 *SV9600 Still Video Transceiver (EKC)*

Our transceiver used a digital image compression technology supplied to us by Majid Rabbani of the Kodak Research Labs. He gave us independently developed technology that was essentially what later became known as *JPEG*. This familiar acronym stands for the Joint Photographic Experts Group, the international standards body that worked to define the eventual standard in 1992. Actually, the compression algorithm in our transceiver was better than JPEG, at least for our application, as it partitioned the image data into 16x16-pixel blocks (JPEG baseline uses 8x8), and also the discrete cosine transform (DSC) coefficients were weighted in accordance with their relative importance based a model of the human visual system (HVS) (also not utilized in JPEG). The HVS normalization was a patented Kodak invention, which was among those that Kodak gave away under free license to JPEG so that all parties to the JPEG standard could agree, and the standard could be born. Majid Rabbani was the first Kodak representative to JPEG, followed by Michael Nier, and they both should have our gratitude for contributing significantly to making the JPEG standard actually happen for the benefit of the world. Images compressed with JPEG technology are ubiquitous in computers and on the Internet today, and use the familiar .jpg to complete the file name. You now use it all the time.

In 1989, CBS News used our *SV9600* transceiver to send some important pictures to New York City during the Tiananmen Square incident, when the Chinese had blocked all video satellite transmissions. The result was some unexpected media exposure for our transceiver, including a 2½-minute special CBS News item broadcast on June 4, 1989, for which the *SV9600 Still Video Transceiver* was given an Emmy Award, albeit a "nerd" Emmy (that is, not in primetime). This first Kodak all-digital product was able to send a color picture over standard phone lines in under sixty seconds with a 9600 bps (bits per second) modem—a big deal back then, as there was no publicly-available Internet. (It's hard to imagine a world without the Internet.)

The *SV9600* model was followed by the *SV9600-S*, which could securely transmit images over encrypted STU-III lines. A third model, the *SV9610*, allowed 64 Kbps image transmission over the new Integrated Services Digital Network (ISDN) lines. In developing the *SV9610*, we worked with the Rochester Telephone Company, and in particular, an excellent engineer named Anita Freeman, who had worked for me at Kodak in Copy Products in the 1970s. This photograph from Rochester Telephone's 1989 annual report shows Don Pophal, our transceiver product manager, with Anita and me.

Fig. 1987.11 *(l to r) Don Pophal, Brad, and Anita Freeman at Rochester Tel (RTel)*

A few years later, Anita left Rochester Telephone to join Pacific Bell, where she became friends and cubemates with a chap named Scott Adams, who based his character Alice in his Dilbert comic strip on her. So, despite the fact that "Alice" once worked for me, I have tried over the years not to be a "pointy-haired boss." Not enough hair, for one thing.

Since we were using the *Still Video System* to show Kodak's prowess in electronics in 1987, we made a publicity picture of my boss, Bill Prezzano, the head of the Photographic Products Group, and me:

Fig. 1987.12
Bill & Brad with some SVS Products (EKC)

Bill was always a supporter of what we trying to do with electronic imaging. He knew it was coming. I enjoyed working for him and learned a lot about finance and business.

All of the *SVS* equipment products were manufactured at Elmgrove Plant in Rochester, and the earnest and capable folks on the assembly line proudly installed "Made in USA" cards in each one and signed them.

Fling Single-Use Camera

In 1987, Kodak also announced its first single-use camera, the *Fling* camera, containing a *110 Kodacolor* film cartridge. This unfortunate trade name was later changed to *FunSaver* to appease environmentalists, and the camera bodies were recycled. This product was being promoted inside the company for some time, but some Kodak executives worried that the product would "cheapen photography." They changed their minds soon after Fuji announced its version. Fortunately, Don Harvey and Dana Wolcott had done the necessary advanced development work, so that Kodak could respond rapidly. Don told me it was really just a "box with a hole in it," but containing the precious film.

Fig. 1987.13
The Fling Camera (GEH)

The Museum of Modern Art included the *Fling* camera as one of the designs selected for its "Humble Masterpieces" exhibit of 2004. Kodak still sells some single-use cameras today (made in China), and historically, they have been a very profitable product line.

Organic Light-Emitting Diodes

Also in 1987, Dr. Ching W. Tang, a senior research associate, and his colleague Steven Van Slyke developed the first multi-layer, organic light-emitting diodes (OLEDs) at the Kodak Research Labs, for which Dr. Tang later became a fellow of the Society for Information Display (SID). These devices are capable of making beautiful, bright, high contrast color-saturated displays. This marvelous technology will come up again later.

1988

Sterling Drug Acquisition

In 1988, Kodak acquired Sterling Drug Inc. for $5.1 billion. This acquisition was intended to provide the infrastructure and marketing ability Kodak needed to be a profitable participant in ethical and over-the-counter drugs. By profitable, Kodak meant that this new product line could potentially make as much money as film.

Now Kodak was in the "pill" business, amongst other things. I was very concerned that this move signified that Kodak was losing faith in the "picture" business, and that this turn of events would be demoralizing to those working in this area, especially in our new field of electronic imaging. That turned out to be correct. Kodak didn't know very much about the pharmaceutical business and how competitive it was because it was not, after all, Kodak's core competency. The Sterling Drug acquisition was ultimately not a success. George Fisher came in as CEO in 1993, and a year later Kodak completed the sale of Sterling and its non-imaging health-related businesses. I don't have the exact numbers, but estimate that the entire Sterling Drug adventure from 1988 to 1994 cost Kodak about $10 billion. Big bucks. Kodak went in debt for this. We will come back to this in the concluding chapter.

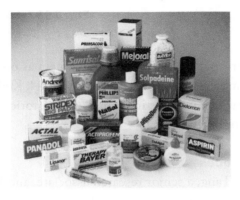

Fig. 1988.1
Sterling Drug Products (EKC)

ColorEdge Copier

In 1988, Kodak became the leader in high-volume color copiers with its *ColorEdge* copier for under $60,000. It was the only American-made color copier on the market and was four times faster than any other color copier at that time. It used the existing copier technology of placing the original on a platen and exposing it to a flash of light. Three years later, in 1991, Kodak introduced the *1500* series that used electronic scanner technology instead of analog optical technology, and in addition, provided stapling, book making, insertion, and folding capabilities.

Qualex

Qualex, Inc. was established as a joint venture, in 1988, between Kodak and Fuqua Industries, Inc., merging the operations of about 90 photographic processing labs owned by the two parties. J.B. Fuqua (pronounced "few-kwa") became rich as he bought and sold over 60 companies involved in sporting goods, lawn and garden equipment, and, amazingly enough, photofinishing. I met J.B. at a PMA conference once, and was favorably impressed with his demeanor, his smooth Southern accent (he was from Atlanta), and especially his very large diamond stickpin in a very bright red tie. He is the founder of the Fuqua Business School at Duke University.

While photofinishing wasn't considered as profitable a business as selling film, it was nevertheless a good business that made money. In addition, it produced what I always claimed Kodak customers really wanted, namely, the actual pictures. So, even though J.B. seemed like a fine fellow, I questioned why we would want to get Fuqua Industries (or anybody else) involved, since photofinishing produced the prints that "touched" Kodak customers.

Polaroid Damage Settlement

Polaroid asked for $5.7 billion in damages after the courts ruled in its favor in 1985 in a suit against Kodak for instant photography patent infringement; in 1988 they got less—$925 million. But that was still a lot of money for Kodak to pay out.

What to Do with Diconix?

The dedicated folk in Dayton, Ohio who were working at Diconix on inkjet printing technology were becoming an odd fit at Kodak, so, as I mentioned earlier, they were absorbed by Kodak Copy Products and given the uninspired name of Kodak Dayton Operations. The troops didn't like the new name because they were proud of the Diconix brand at that point, and, I felt, rightly so. I didn't know it yet, but they were all about to be working for me.

Create-A-Print

The *Kodak Create-A-Print 35 mm Enlargement Center* enabled consumers to crop and print their own enlargements in a few minutes. This original version took a strip of 35 mm film, scanned the negatives, showed positive images on a screen and made prints with a semi-dry processor. It was a tough sell inside Kodak to get this product launched, eventually requiring us to build four different prototypes. The bright and inventive Scott Brownstein at the Kodak Research Labs developed the basic idea and built the first working prototype, which I recall took in a 35 mm film strip and made an instant print. Scott was a trusted colleague and he persisted despite a negative view of the prototype. One marketing opinion was that "no one was going to stand in line to make pictures," and another repeated the refrain I had heard for years, that the image quality from the kiosks "wasn't *Ektachrome*." Eventually, the kiosk business became very successful for Kodak, with over 100,000 kiosks worldwide. It was sold off in 2013.

Slide/Video Transfer Unit

The *SV5035 Slide/Video Transfer Unit* was announced in 1988. It took an 80-slide *Carousel* slide tray, and converted the slide images to video. Commercial customers had requested this capability when the *SVS* line of products came out in 1987.

Fig. 1988.2 *SV5035 Slide/Video Transfer Unit (EKC)*

John Acello worked on this and incorporated many useful features, including random access, motorized zoom and scan, and rotation. This was another example of a hybrid product that made constructive use of film technology as well as electronics.

Prism XLC Electronic Previewing System

Following on from work done on the *SV8300 Still Video Camera* and the *Kodak Video ID System* (*KVIDS*), Tom Nutting and his team developed a

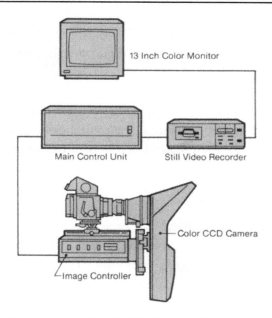

Fig. 1988.3
The Prism XLC Electronic Previewing System (EKC)

great hybrid product for the Professional Photography Division that allowed proofing traditional film photography in the studio in real time. The drawing above, from a Kodak brochure called the "Studio of Tomorrow," gives the idea.

The professional film camera sat on top of the image controller and was aimed at the subject; simultaneously, a color CCD camera saw the same image reflected by an optical prism, and that electronic image was sent to a color monitor. The image controller synchronized the flash image on both cameras, so the monitor image was the same image that was being recorded on the film, except that it was viewable immediately. The *Still Video Recorder* could record all the images taken during the shoot, to be reviewed by the client before leaving the studio.

Previously, instant prints were often used to preview film prints in studio photography, so that the client wouldn't have to wait days to obtain proofs printed out by the studio. The problem with these paper proofs was that customers would take them home, and then, as the thrill wore off, would often place only small orders (and keep the proofs). The *Prism* system, on the other hand, enabled customers to order prints on the spot, knowing what they would look like; it was more exciting, and the professional photographers told us that when they used the system, customers tended to order more prints. Photographers also liked it because it helped them do a better job of posing and lighting, so their portrait quality improved.

Many studios adopted this system that sold through the 1990s. In 1988, however, few people realized that one day in the future most professional studio photographers would be using just a digital camera and no film! All hybrid products were eventually doomed to bite the dust, of course. Hybrid gas/electronic cars are headed for the scrap heap when they go all electronic.

Kodak/Hitachi Technology Agreement

During the previous two years, I had traveled to Hitachi headquarters in Japan to discuss ways we could cooperate in consumer electronics, because Kodak was still concerned that we couldn't compete with the Japanese in electronic equipment. I fundamentally disagreed with this notion, but at the same time found the people at Hitachi to be excellent engineers and managers, and along the way we worked out a deal where they would make a thermal printer for the consumer video market and Kodak would supply the thermal media. Kodak management was very happy with this outcome, because it felt more comfortable with "media" as opposed to equipment. This arrangement was announced in 1988, and the products came out in 1989. Our thermal media was co-branded in Japan but just branded as a Kodak product in the U.S.

Tactical Camera

Often, advanced technologies are first developed for the government, and especially the military, as with spy satellites in the 1960s. This was also the case with the earliest digital single-lens reflex (DSLR) cameras that were designed in the Kodak Federal Systems Division (FSD) and later in the Professional Photography Division (PPD) by Jim McGarvey and others.

Since Kodak had come up with the world's first mega-pixel CCD in 1986, one of FSD's governmental agency customers asked whether a digital camera could be made for its own use, resembling the *Canon SLR* cameras it had worked with previously, so that its employees could use the same lenses.

Fig. 1988.4 *Tactical Camera (GEH)*

Jim designed the first DSLR in 1987, called it the "Electro-Optic Camera," and produced one for the agency. In Jim's own words, talking to Todd Gustavson, curator of technology at George Eastman House:

> *That agency took the camera, and I've never seen it since. After that, we took that basic design, repackaged it a little bit, and made two of what we named 'Tactical Cameras.' They are the second and third DSLRs ever created, and one of them is what you've just acquired.*[7]

He was referring to the fact that one of these two *Tactical Cameras* was found in 2012, and given, happily, to George Eastman House. Of this acquisition, Todd Gustavson said,

> *The Tactical Camera may well be the most important object acquired during my 24 years at Eastman House. There is nothing like it in the collection, but more importantly, it is one of only two of its kind ever made, and it is from these models that all digital cameras were derived.*[8]

He was right. At Kodak, these models evolved into the *Professional Digital Camera System* that launched in 1991, at the dawn of high-quality digital photography, developed by Kodak.

1989

Digital Continuous Tone Printer

While Kodak had good success with our *SV6500* thermal printer that made 4x5-inch prints, customers almost immediately asked for bigger and better prints. So we developed the *Kodak XL7700 Digital Continuous Tone Printer*, which produced large-format thermal color prints (up to 11x11 inches in size) and transparencies (at up to 200 pixels per inch). Now that computers were able to display high-quality color images, this innovative product provided a way to make high-quality prints and transparencies to use on overhead projectors. (Remember overhead projectors?)

Fig. 1989.1 *XL7700 Digital Continuous Tone Printer (KBP)*

A critical impetus for this printer came from the Federal Systems Division, run by one of my colleagues from the Kodak Apparatus Division (KAD) research lab, Gary Conners. He said he had ready government clients for such a printer—something odd like printing maps inside trucks in the desert. So, in uncharacteristic Kodak business unit fashion for the time, we collaborated and shared people and budgets to design and build this printer. This partially accounts for the rack-mountable, industrial-strength design. Scott McNealy, CEO of Sun Microsystems, which was partnered with Kodak at the time, loved these printers, because he had just developed a line of Unix-based graphic workstations, and now he could get high-quality digital prints and transparencies from them. When, after a demo and during a break, I told him I had resorted to bootlegging the tools to get the *XL7700* into production, he was tickled, and told me, "To ask permission is to seek denial." I have repeated that quote many times over the years, and still believe it's true.

The grandchild of this printer (*ColorEase*, 1993) was smaller and less expensive, even though it offered greater resolution (300 pixels per inch). It was used with great success in later models of the original *Create-A-Print* kiosks (later called *Kodak Picture Makers*). We intentionally kept the thermal print group intact for long enough to get really good at thermal print technology ... more "core competency" at work.

More One-Time-Use Cameras

Kodak one-time-use cameras were a great success, and two new models were added for special purposes. The one-time-use *Kodak Stretch 35* camera produced 3½ x 10-inch prints for panoramic scenes. In addition, the one-time-use *Kodak Weekend 35* camera was an all-weather camera capable of taking pictures underwater to a depth of 8 feet. Neither of these models were huge sellers, but they added richness to the product line, an appropriate goal for a company committed to the picture business.

End of the Eighties

At the beginning of the 1980s, Kodak had no debt on its balance sheet. That began to change during the following years, and in 1988 it radically changed with the $5.1 billion purchase of Sterling Drug. At the end of 1989, Kodak's long-term liabilities totaled almost $9 billion.

In December 1989, Colby Chandler announced that he would step down in 1990 when he turned 65. Kay Whitmore, architect of the Sterling Drug acquisition, would be his replacement. Phil Samper, vice chairman of Kodak at the time, was my personal pick to replace Chandler. I liked him because he was interested in what we were doing, stayed awake at our briefings, and asked good questions. On the very day Whitmore was chosen, Samper announced that he would retire on January 1, 1990. Phil had several important jobs after leaving Kodak, including being the president of Sun Microsystems.

So by the time Kodak reached the end of the 1980s, it was involved in (at least) three of what seemed to me rather disparate businesses: pictures, pop bottles and pills. I sensed nervousness in the ranks that the future was becoming less clear. I was starting to worry that more layoffs would come, and, being basically an R&D guy, I was worried that I would have to lay lots of people off in the future, not my cup of tea.

5

THE NINETIES

1990

Photo CD System

In 1990, Kodak announced its *Photo CD System* for storing images and playing them on television and computer screens, and proposed this as a worldwide standard for defining color in the digital environment of computers and computer peripherals. The overall system (see next page) was a hybrid design like its analog predecessor, the *Still Video System*.

The hybrid nature (film and electronics) of *Photo CD* was emphasized in many advertising photos, like the one below:

Fig. 1990.1 *Photo CD & Film Hybrid Ad (EKC)*

Here you see a strip of film, which can be scanned and recorded on a *Photo CD*, and in fact, a great early use of *Photo CD* was just to get high-quality, low-cost film scans for computer use. You also see a jewel case for CD storage, and an index print, made by a variant of the *Kodak XL7700 Digital Continuous Tone Printer*, to identify what pictures were on the CD.

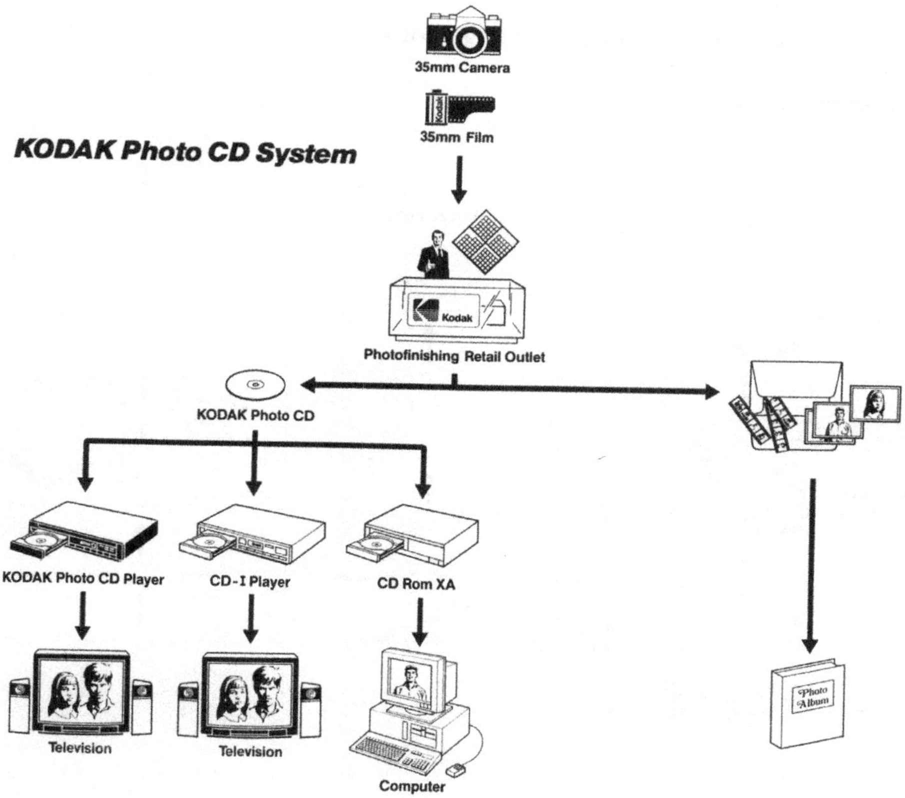

Fig. 1990.2 *Photo CD System (EKC)*

The *Advantix* photo system adopted this index print idea later in 1996.

One reason for the popularity of *Photo CD* for commercial use, especially for prepress and film scans, was that the image on a *Photo CD* was stored in five resolutions, the highest (16-base) being 2048×3072 pixels. This is about 6.3 megapixels, and is enough to completely capture the image from most 35 mm film frames. Frank Cost wrote a lovely small book while at the RIT Research Corporation in 1993, entitled *Using Photo CD for Desktop Prepress*,[9] that explains all this more comprehensively.

If you were really a pro and had a very high-end camera, perfectly in focus, perfectly still during exposure, with slow speed film, no dust on the optics, and great lighting, then you may have opted for the *Pro Photo CD*, which allowed a 64-base capture of 4096×6144 pixels, or about 25.2 megapixels. This system was mostly used for very big enlargements.

Here is a more complete schematic of the hybrid *Photo CD System*:

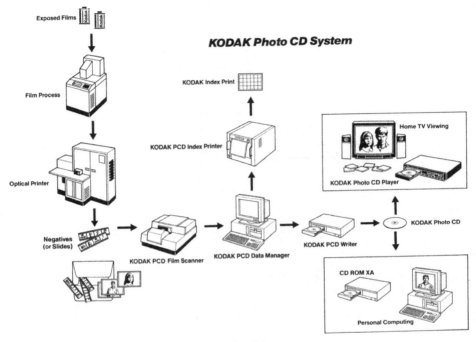

Fig. 1990.3 *Hybrid Photo CD System (EKC)*

In an unfortunate bid for premature consumer adoption of the *Photo CD System*, Kodak came out with a *Photo CD Player*, as seen here:

Fig. 1990.4
Photo CD Player (EKC)

It seemed like a nice idea, and some fancy travel books that Kodak sponsored started including a *Photo CD* in an envelope inside the rear cover. But the *Photo CD Player* never caught on due to its price, its limited usefulness (it only played *Photo CD*s), and because the *Photo CD* data formatting standard was kept proprietary by Kodak.

Again, as was the case for the *Still Video System*, Kodak also offered a *Photo CD Transfer Station* (below) for labs to scan film to the *CD*.

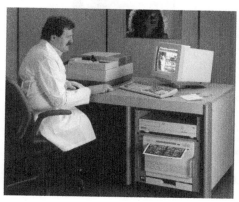

Fig. 1990.5
Photo CD Transfer Station (EKC)

Internal Kodak attempts to market *Photo CD* as a "digital negative" (which it actually was, by design, from a technical standpoint) were squashed by Kodak management because "It might hurt film." The *Photo CD System* was intended by its internal Kodak developers to be aimed first at commercial markets, but Kodak management decided instead to market it to consumers first, which, as I suggested above, was a major error.

Adobe Photoshop

The same year, Adobe released *Photoshop 1.0* for Macintosh computers, enabling digital picture enhancement and manipulation. *Photoshop 1.0* sold for $1,000 in 1990, and by the year 2000, three million copies had been sold. Not bad. I always thought that this should have been a Kodak product—we had been working on our own "digital darkroom" but found out that the name was being used by somebody else. We were discouraged from proceeding by being told that we weren't in the software business. However, Adobe certainly was in the software business, having developed *PostScript*. Leaping ahead a decade and quoting from Adobe's 2011 annual report: "Throughout fiscal 2011, we maintained our focus on making *Photoshop* the standard by which all other imaging products are measured." Adobe showed constancy of purpose and succeeded.

Fig. 1990.6 *Adobe Photoshop 1.0 (CHM)*

Hubble Trouble

In April 1990, the renowned *Hubble Space Telescope* was launched and placed into orbit by the space shuttle *Discovery*. The initial pictures were blurry, and it was soon discovered that the primary mirror had been ground and polished incorrectly by Perkin-Elmer (P-E); it had an optical defect known as spherical aberration. In fact, there was a back-up mirror, correctly made by Kodak, in case there was a crack or defect in the P-E mirror, but since the problem wasn't discovered until after launch (due to poor testing), it was too late to use it. This back-up mirror now resides in the National Aeronautics and Space Museum, as shown on the next page.

You can see the light-weight honeycomb construction through the front glass because the reflective coating was never put on. Since there are no scale clues in the picture, I should tell you that it is almost eight feet in diameter! Bringing the *Hubble* back to earth for a retrofit was out of the question.

NASA formed a special commission headed by Lew Allen, then the director of the Jet Propulsion Lab (the division of NASA that helped reconstruct *Lunar Orbiter* photos), to see what could be done. Chuck Spoelhof, my former boss from the spy satellite days, was on that commission, and he helped them figure out a way to fix the problem. The commission suggested inserting two (small) mirrors in the optical path, one of them ground and polished to correct the spherical aberration in the primary mirror.

The *Hubble*'s first servicing mission flew aboard the space shuttle *Endeavor* in December 1993, installed the corrective optics, and NASA declared the resulting pictures a success. Now the world had a fully capable space telescope, no small thanks to Kodak.

Fig. 1990.7 *The Hubble Back-up Mirror Made by Kodak (NASA)*

1991

Professional Digital Camera System

The initial groundbreaking work begun by Jim McGarvey with the *Tactical Camera* in 1988 continued at Kodak, and the *Kodak Professional Digital Camera System* (*DCS*) was introduced in 1991. It enabled photojournalists to take digital pictures with a Nikon F3 camera body equipped by Kodak with a 1024x1280 (1.3 megapixel) sensor. The first model, unofficially called the *DCS 100*, was the first commercially-available, single-lens reflex digital camera, 16 years after the first digital camera had been invented by Kodak in 1975. It was bulky by today's standards, of course, and had an over-the-shoulder *Digital Storage Unit* to both store and enable photographers to view the images taken. It initially sold for over $20,000, a tad steep for home use, but almost a thousand photojournalists bought one.

Fig. 1991.1 *DCS 100 Photojournalist Camera (GEH)*

The cameras came to have a nice logo, too:

Fig. 1991.2 *DCS Logo (EKC)*

There were many, many *DCS* models that came after the *DCS 100*. Ken Parulski of KRL was involved in all of them and told me he plans to write a book on electronic cameras that will cover them in more detail.

DCS cameras continued to be sold through the early 2000s, but then the *DCS* line was discontinued due to the fear of hurting film, a concern over investing in marketing and R&D, and an excessive focus on near-term financial results.

Approval Proofing System

The *Approval Digital Imaging System*, designed for prepress proofing, was introduced in 1991, and is still sold today.

Approval was an innovative outgrowth from the Kodak Research Labs of our thermal printing technology, but in the *Approval System*, the heating

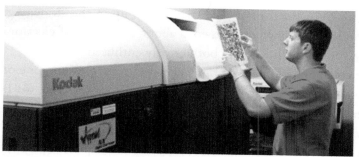

Fig. 1991.3 *Approval Proofing System (EKC)*

of each dye pixel to transfer the dye was produced not by a fax head, but by a laser—a notion developed by Chuck DeBoer. The prototypes made absolutely great continuous-tone prints. Nevertheless, Hal Gaffin, who ran the Graphics Division, told me one day that most traditional printing customers wanted halftone dots for proofs; never mind that, as I said, the quality of the continuous-tone prints was up to the task of proofing!

I talked with the team in research, asking, "You can make dots with this thing, can't you?" They said, "Yes, but that would be underutilizing the technology." I understood what they meant; dots are conventionally used in printing to approximate the quality of continuous tone photography. But I explained that if traditional printing customers wanted dots, maybe it didn't matter if we thought they needed them or not, dots were part of their requirement. So we began to make dots!

Not that it was simple, because the *Approval* system gives control over screen angles, screen ruling, density control per color, dot gain adjustment and dot shapes. My friend and colleague Ed Granger was the technical assistant to Hal Gaffin back then, and was instrumental in getting the Graphics Division to accept this new technology. *Approval* was a lovely system, and it started to replace the previous proofing system, *Signature*, based on liquid toner electrophotography.

Technology Drivers

I worked on and off at Kodak with a great guy, Dave Lehman, who came originally from IBM and then left that company to join Jack Goldfrank from Xerox in forming Mead Digital Systems, which later became Diconix when Kodak bought it in 1983. Dave wound up in the strategy group at Kodak, and we had many discussions about the future of electronic photography, about which we were both greatly enthused.

We collaborated on several management presentations, one of which foretold the significance of the three "technology drivers" we believed would enable electronic photography to become a successful enterprise: memory cost, processor power, and communications bandwidth. In one of our slides, below, memory cost is measured in dollars per megabyte, processor power in millions of instructions per second, and communications bandwidth in millions of bits per second.

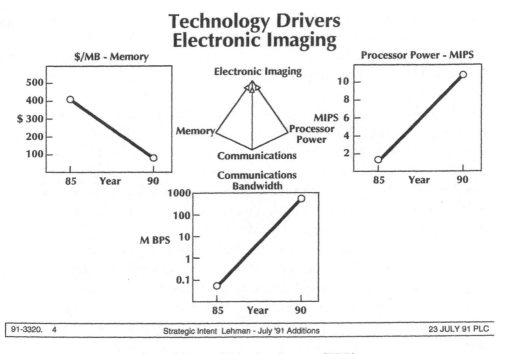

Fig. 1991.4 *Technology Drivers (KBP)*

The idea was that as technology marches on, computer memory costs drop, computer-processing speed increases, and communications speed increases. These parameters would be the drivers for electronic imaging to be of increasing practical use. This turned out to be an accurate conceptual model, but it didn't seem to ring the urgency bells for Kodak at that time.

A Value Model for Cameras

Along with this technology driver notion, I had been putting forth a camera *Value* model for several years, where I defined *Value* (*V*) as:

$$V = Q / C$$

Here, *Q* = quality, defined as lines of resolution per picture height, and *C* = cost, defined as the number of dollars required to capture an image and be able to see it, either in a viewfinder or from an available printer. ("Sparkies" will also recognize this as the famous formula for voltage equals charge over capacitance, but I digress.)

So, for example, a simple single-use film camera at the time had a terrific *Value* of more than 100, because it captured images with at least 1,000 lines of resolution per picture height and could be purchased for $10. A box 35 mm camera could be bought for $100 and yielded pictures with upwards of 2,000 vertical lines of resolution, and so in this case the *Value* was 20. I hypothesized in the late 1980s that when, say, an electronic camera selling for $1,000 could produce 1,000 lines of resolution per picture height (that is, *V* = 1.0), the digital electronic imaging era would soon be upon us, like it or not. Admittedly, this was a simple heuristic model, but turns out it had pretty good predictive ability.

News Flash: Customers are pretty smart when it comes to assessing true value. How many single-use cameras were sold in the U.S. in the ten years between 1997 and 2006? Answer: 1.7 billion! Yes, that's billion!

As a digital camera example of this simple but effective *Value* model, consider the *DCS 100* described above. It sold for $13,000 and had 1024 rows of pixels vertically, so its *Value* (*V* = *Q* / *C* = 1024/13000 = 0.08) was quite a bit less than 1.0. So, you may ask, when did digital cameras get to 1.0, and when did they get to 20? Remember *V* = *Q* / *C* and read on.

1992

The Big Buyout

As mentioned in Chapter 1, I took the "3R" retirement buyout from Kodak after 32 years of working on various electronic and advanced imaging projects. After retiring I continued to have lunch with former colleagues, but quit this practice after a few months because it became too depressing. I enjoyed consulting work, including ten years of patent expert litigation activity, and never looked back.

More New Digital Products

New digital products in 1992 included the *Kodak Professional DCS 200 Digital Camera* and the *Kodak XLT 7720 Digital Continuous Tone Printer*.

1993

Portable Photo CD Player

The following year, Kodak launched several new software products and *Photo CD* formats for commercial use, including a portable *Photo CD Player*. It was cute, and it worked, but it didn't sell well.

Cineon

Kodak technicians digitally restored Walt Disney's 1937 classic, *Snow White and the Seven Dwarfs*, using a new technology, *Cineon*, which had been in development inside the Kodak Research Labs for some time. It was the first system to successfully scan film, digitally process the scanned images, and print the processed images out to film in volumes needed for enhanced-quality movies. The technology could restore faded colors and remove dirt and scratches.

Fig. 1993.1 *Cineon Movie Processing System (EKC)*

Cineon used some pretty serious computing power for its time, and heralded all the digital enhancement and special effects so prevalent today. This basic technology, producing what movie folk call a digital intermediate, is now in use worldwide.

Just before I left Kodak in 1991, I spoke with some of my research lab colleagues about the prospects of digitally enhancing *Snow White*, realizing that we would need to store more than a terabyte of data! This caused quite a bit of commotion at the time, because a terabyte was rather unheard of, except, of course, in the labs. One of our optical disc jukeboxes, fully loaded, could hold a little over a terabyte. (A terabyte is a thousand gigabytes, and a gigabyte is a thousand megabytes. In the Apple store in late 2012, I saw that you could buy a two-terabyte drive for under $200. Blew my mind.)

Imagelink Scanner

Kodak introduced a hybrid scanner that could not only image documents onto microfilm but also scan them at the same time. It was called the *Imagelink 990D Scanner*, shown below:

Fig. 1993.2 *Imagelink 990D Hybrid Scanner (EKC)*

This scanner has a warm spot in my heart, largely from the point of view of a user. At this time (1993), I was working at the RIT Research Corporation on projects for the U.S. Census Bureau, an activity that continues to this day at ADI, LLC. The Bureau was already familiar with the previous microfilmers from Kodak, because it had used them to microfilm all of its Census forms, and had even built its own "page-turning cameras" that would microfilm forms that were booklets. So, when we wanted to encourage the Census Bureau to get into electronic imaging for processing the upcoming 2000 Decennial Census, starting our development work with the *990D* scanner was a natural. We had one at RIT and there was one at Census headquarters in Suitland, MD, and off we went. (This is a long story, which is detailed in my book *Handprint Data Capture*.)

Diconix Sold

Also in 1993, Kodak sold Dayton Operations (Diconix) to Scitex for $70 million. I was really, really unhappy, as I have said previously. Big mistake. Then in 2004, amazingly enough, Kodak bought it back for over three times as much as it was sold for, and of course was quite proud in the news release of its intelligent digital printing purchase. I guess I can understand why Kodak didn't mention the name Diconix in the news release, but nevertheless, it didn't seem proper.

Kodak Picture Exchange

One of the burning questions at Kodak always was, "How are we going to make money on this electronic stuff?" (Translated: "How will we make as much with electronics as with film?") One idea was *Kodak Picture Exchange* (*KPX*), which was launched in 1993 with photos from 14 stock photo agencies, each supplying between 5,000 and 20,000 images. It was sold as software that provided access to professional-quality images, with Kodak taking a little piece for the trouble ($9/proof). It was an innovative idea, suggesting a Kodak vision for a global image transmission services network. The *KPX* system would display an image like the example below, where the search topic was "pets."

Fig. 1993.3 *Kodak Picture Exchange (EKC)*

ColorEase Printer

Kodak introduced a family of small-profile thermal printers capable of making 8x10-inch prints and transparencies at 300 pixels per inch resolution, culminating in the *ColorEase PS Printer, Model 8670*:

Fig. 1993.4 *ColorEase PS Printer (SS)*

The *PS* stood for *PostScript,* the printing font language developed by Adobe. These printers were an outgrowth of the larger *XL7700* printers (which used 200 pixels per inch resolution), but had a roller configuration instead of a drum and so were smaller. They were incorporated into the *Picture Maker Kiosks* the following year (1994). The project manager was Don Pophal, and Steve Sasson was chief engineer. Gary Lum developed the donor media, which included a clear lamination layer in addition to cyan, magenta and yellow donor layers, giving the resultant prints a durable finish. It was a great product.

Eastman Chemical Spun Off

I recall some meetings with people from Eastman Chemical (always called by us Yankees "Tennessee Eastman") to attempt to find "synergy" or something between the Kingsport and the Rochester contingents. As mentioned previously, I never thought that these meetings were worthwhile, but I did clearly pick up the idea that Eastman Chemical wished to be out from under the boot of Kodak on more than one occasion. As Emily Plishner wrote in *Chemical Week,* "Seventy-three years after George Eastman established Eastman Kodak's chemical affiliate, Kodak CEO Kay R. Whitmore, under intense pressure from investors, would free Eastman Chemical to realize value for shareholders independently."[10] The article continued: "'I've been waiting for them to do this for years,' says analyst Brenda Lee Landry of Morgan Stanley. 'Get this asset in the hands of the shareholders.'"

At year-end, the Eastman Chemical Company (including the Distillation Products business that I recall made Vitamin D), founded in 1920, was spun off to shareholders and became an independent company with its own board of directors and New York Stock Exchange listing. Kodak was now out of the "pop bottle" business. The move did not directly realize cash for the parent, but Eastman Chemical took on about $2 billion of Kodak's nearly $10 billion debt load. Eastman Chemical Company is now a Fortune 500 Company. By itself.

1994

Sterling Drug Sold

George Fisher was now CEO, and Kodak divested its non-imaging health-related businesses—Sterling Winthrop, L&F Products and Clinical Diagnostics—"enabling the company to focus all of its resources on its core imaging business," as the press release stated.[11] Kodak was no longer in the "pill" business, after a period of only six years. Proceeds from the sale of these businesses were used to substantially reduce Kodak debt.

Even though I was long retired from Kodak by now, I was somewhat encouraged by this, and thought that maybe George Fisher was just the guy to turn Kodak around, because, as I always had said, Kodak was in the picture business.

Associated Press Photojournalist Camera

My colleague Jerry Magee was product manager for the *DCS Camera* and deeply involved in some of the good marketing work that was done with the *DCS* line. He told me that they began to work with the Associated Press (AP), and developed a news camera called the *NC 2000* in 1994, which was first sold to AP member newspapers for $17,500. It became the de facto digital news camera for a time, and about 550 cameras were produced. Here is a picture of the *NC 2000e*, one of four models produced for the AP, and as you can see, it was considerably more compact than the *DCS 100* from 1991.

Fig. 1994.1
NC 2000e AP Photojournalist Camera (GEH)

Jerry also told me that one of the problems with marketing the *DCS* cameras internally to Kodak was that a large mark-up was required to make the cameras closer to "acceptable," relative to film profits. Many executives were also afraid to "hurt film" by being overly aggressive with promoting ever-better digital cameras.

Apple Quicktake Camera

The *Apple Quicktake 100* digital electronic camera was launched at the Tokyo MacWorld Expo in 1994. It had actually been designed by Kodak engineers to work with Apple computers, a fact not widely known because there is no Kodak logo on the camera, by design. The *Quicktake* got widespread attention; they called it a "computer camera." It could store eight 640×480-pixel images, or thirty-two 320×240-pixel images, in any of the three popular image formats: TIFF, PICT, or JPEG. It was initially priced at $749, and appealed to many early adopters.

Fig. 1994.2
Apple Quicktake 100 (GEH)

So, in hi-res mode, the *Quicktake* had a *Value* of V = 480 lines/\$749 = 0.64. Hmmm, the *Value* was starting to approach 1.0! Stay tuned.

Picture Maker Kiosk

In 1994, Kodak also launched the *Picture Maker Kiosk*, which allowed digital prints to be made from either conventional photo prints or from a variety of digital inputs. The kiosk used a version of the *ColorEase* thermal printer, as mentioned above. Thermal print technology is a perfect fit for kiosk use not only because the quality is terrific, but also because it uses no liquids, has no print head nozzles to dry up, and can just sit there patiently waiting for the next time a customer would press the Print button. Prior to *Create-A-Print* (1988), we had developed four different prototypes before achieving serious internal Kodak interest. There were over 100,000 kiosks in place worldwide as of 2012. In 2013 Kodak put this business on the block; a desperate move, in my opinion.

Fuji Sponsors the Olympics

Fuji sponsored the Los Angeles Olympic Games for the first time in 1994. This caused considerable "agita" at Kodak, as it clearly symbolized that Fuji was coming after Kodak, big time. Kodak was no longer free to charge as much for film as it had been doing, now that there was a formidable competitor around. Although Fuji film was heavy on bright, saturated colors, and internal Kodak purists dismissed them as "unrealistic," many people liked them, and of course, the lower price was attractive as well. My wife, Joyce, has never quite forgiven me for not buying some Fuji film to try, because she saw other people's prints and liked the colors. My (persistent) true-yellow blood wouldn't allow it, and we never did try it.

1995
Point-and-Shoot Digital Camera

In March 1995, Kodak made news in the digital imaging business with the release of the *Kodak DC40 Point-and-Shoot Digital Camera*. The *DC40* allowed you to capture images and then quickly transfer them to your computer to save or manipulate them using image-editing software. The camera was very easy to use, and offered the following features:

Fig. 1995.1 *DC40 Point-and-Shoot Camera (EKC)*

- Flash settings
- Timer settings
- Exposure settings
- Energy-saving sleep mode
- Built-in lens cover
- Choice of battery or power adapter as power supply
- Battery use display
- Capability for accessory lens attachments

The *DC40* had a 504×756-pixel CCD, and sold for $960, so $V = 0.53$, slightly less *Value* than the *Quicktake*, but with some nice customer-friendly features. Not by accident, it also resembled the *Quicktake*, as you can see. Ken Parulski of KRL, who was heavily involved in the *Quicktake*, told me that the internal electronics of the *Quicktake* and the *DC40* were essentially the same.

The *DC40* was the first Kodak-branded consumer digital camera.

Copier Sales and Service by Danka

In September of the same year, Kodak announced that Danka Business Systems PLC would sell and service Kodak high-volume copiers throughout the U.S. and Canada. Kodak was never wildly enthusiastic about the copier business, partly because of the up-front cost to manufacture copiers, which were then leased to customers as was done at Xerox, the return on the investment being very gradual. This announcement was not, obviously, a good sign for the copier business at Kodak, and we all knew it.

1996

Spirit DataCine

Development work on high-speed and high-quality film scanners had been going on at Kodak throughout the 1980s and continued into the 1990s. The previously-mentioned *Cineon* system, launched in 1993, had some drawbacks, including pin registration of the film frames and a closed software system that did not appeal to Hollywood. The *Spirit DataCine* was launched in 1996 as a joint effort of Philips and Eastman Kodak. Kodak supplied the film scanner, which was welcomed by Philips because it was the best available at that time.

Within Kodak, the project had been called Sandcastle, and my friend and colleague Art Cosgrove came out of the disk drive business in the Mass Memory Division at the end of 1987 to join the project as technical HDTV project leader, working with Dave Lewis in the electronics research lab in KRL. Using a xenon intense white light source along with Kodak-designed linear CCD sensors and a continuous film transport, the product's quality and speed development progressed rapidly. The project also involved Roger Lees from Kodak Limited in England, and Andy Kurtz, Les Moore, and Herb Erhardt from KRL.

The *Spirit DataCine* launch was in April 1996 at the National Association of Broadcasters (NAB) convention, and the product was chosen as the NAB Editor's Pick of Show. Worldwide, over 370 systems were sold (at over $1 million each) and are still used today in postproduction facilities to convert movie film productions to NTSC, PAL, or HDTV formats. When you watch movies on HDTV that originally were captured on film, you are enjoying the results of this fine telecine product that Kodak had a major role in creating.

Advanced Photo System

Kodak's *Advanced Photo System* (*APS*) format was introduced in 1996, using the *Advantix* brand name. Features included drop-in film cartridge loading, mid-roll change enabling the film to be removed before being completely exposed, and three different picture formats—classic, group (high definition TV aspect ratio), and panoramic. The *APS* format could also add magnetic metadata to the film, like time and date for back printing. I never got too

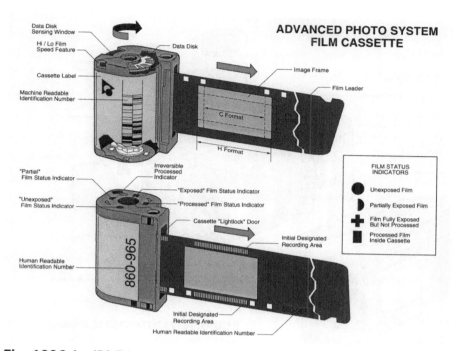

Fig. 1996.1 *APS Film Cassette (U of R)*

excited about the supposed "advanced" features of all this, but the cartridge load was slick. The film cassette is shown above.

The *APS* film cassette was very easy to load, and there was a handy book-like container that made it convenient to store cassettes for posterity. The camera took the same picture (*H*) regardless of what format it was in—the selected format only affected the print. Specifically, the formats were:

- *H* or high definition (30.2x16.7 mm; aspect ratio 16:9; 4x7-inch print)
- *C* or classic (25.1x16.7 mm; aspect ratio 3:2; 4x6-inch print)
- *P* or panoramic (30.2x9.5 mm; aspect ratio 3:1; 4x11-inch print)

Don Harvey had invented the format idea for a panoramic single-use camera. He had called it "pseudo-pan" because you just used a small slice of the available film to make the print. This worked if you didn't need big enlargements. In this case, the *APS* film was 24 mm wide to start with, giving about half of a normal 35 mm film area, but it still made good pictures.

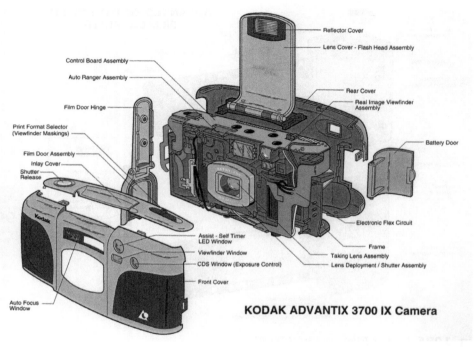

Fig. 1996.2 *Advantix 3700 IX Camera (U of R)*

The camera had a handy, slick design as shown in the exploded view above. The flip-up cover helped reduce red-eye by moving the flash away from the lens. The camera had auto-focus exposure control, and was a cinch to load. Because the first, roughly $100 million, launch of the *APS* line faltered due to the lack of cameras available in stores and the lack of processors in labs to develop the images, there was a subsequent, roughly $100 million, "re-launch." Overall, the camera didn't offer enough perceived advancements to be a commercial success, and suffered the additional indignity of having to compete with the new digital cameras that were starting to be interesting, if not yet widely popular. Kodak ceased *APS* camera production in 2004, and discontinued *APS* film production in 2011.

1997

Danka Acquires Kodak's Copier Business

Kodak sold the sales, marketing, and equipment service operations of its

Office Imaging business and its facilities management business (formerly known as Kodak Imaging Services) to Danka Business Systems PLC in 1997. Even though Kodak kept the manufacturing business, I believe this was the death knell of the copier business at Kodak.

Kodak Picture Network

The *Kodak Picture Network* was announced the same year, enabling people to view photos, order reprints, and share pictures with friends and family around the world via the Internet. This was an exciting and promising new aspect of the picture business.

Digital Science Zoom Digital Camera

In April, the company unveiled the *Kodak Digital Science DC120 Zoom Digital Camera*, the first point-and-shoot, megapixel-quality digital camera under $1,000. Kodak was now seriously into the consumer digital camera business, finally, over twenty years since the invention of the first digital camera by Steve Sasson and Gareth Lloyd at Kodak in 1975.

Fig. 1997.1 *DC120 Zoom with a Megapixel! (EKC)*

This camera had a 960×1280-pixel CCD, so its *Value* was roughly $V = 960/1000 = 0.96$, or almost one! The photographic digital age was coming on fast now, as predicted by my simple *Value* model.

Wang Software Acquired

In 1997, Eastman Kodak Company also finalized the purchase of Wang's software business unit, called Eastman Software, Inc., and put it in Kodak's

Business Imaging Systems organization. The acquisition promised to give Kodak valuable workflow, imaging, document management, and storage management software capabilities.

1998

You've Got Pictures!

In 1998, America Online and Kodak announced *You've Got Pictures!*—a service whereby AOL members could request their processed pictures to be delivered online. No doubt it seemed like a good idea at the time, but it didn't fly.

Kodak Professional DCS Cameras

The *DCS* team led by Jim McGarvey continued to bring out new *DCS* cameras, and a partnership with Canon led to the *Kodak Professional DCS* line of cameras in 1998. Here's a lovely camera that Jerry Magee, product manager for the *DCS* line, colorfully called a "kick-ass" camera, the *Kodak Professional DCS 520*:

Fig. 1998.1 *Kodak Professional DCS 520 (GEH)*

It was 1152 x 1728 pixels, or almost two megapixels, had a compact design, a Bayer RGB color filter array, and came out at PMA for $14,995.

1999

Copier Manufacturing Sold to Heidelberg

The following year, Kodak sold its digital printer, copier/duplicator, and roller assembly operations to Heidelberger Druckmaschinen AG. The two companies also expanded their joint venture, *NexPress*, which had been created in 1998 for high-speed duplicating. So now, despite hanging on for 24 years, the copier business at Kodak was totally gone.

First OLED Display
Kodak and Sanyo Electric Co. unveiled the world's first commercial model of a full-color, active-matrix, organic light-emitting diode (OLED) display.

End of the Nineties
With George Fisher now in charge, it seemed that Kodak had renewed its commitment to being in the picture business. The 1998 annual report (published in 1999) had a business definition floating in the clouds that gave a little more detail:

> **Our Business Is Pictures**
>
> ▶ For Memories As long as people want pictures of their best memories, and as long as people look for new ways to use pictures – we can grow our business.
> ▶ For Information As long as people need information – on a space mission, a food trend, or a health exam – Kodak will find new commercial uses for pictures.
> ▶ For Entertainment As long as people are stimulated by movies, TV shows, or theme parks – Kodak can expand the market for new entertainment experiences.

Fig. 1999.1 *Annual Report Business Definition (EKC)*

Although I was long absent from Kodak by 1999, I thought that the "Our Business Is Pictures" statement was excellent, concisely covering consumer products (memories), commercial products (information), and movies (entertainment).

Thus ended the 1990s, and Kodak was in trouble, despite rosy projections from the CEO's office. From 1998–1999, Kodak's stock price decreased from $72/share to $66/share, and worldwide Kodak employment dropped from 86,200 to 80,650. Dan Carp took over as CEO in 2000, and although Kodak shipped a lot of digital camera models in the start of the new century, the company lost money on most of them. Customer-perceived value for digital pictures, however, was steadily on the rise.

6

THE TWO THOUSANDS

2000

Phogenix Joint Venture with Hewlett-Packard

In the late 1990s, HP considered buying Kodak. Subsequently, Antonio Perez, who was a top executive at HP, was behind the creation of Phogenix Imaging, LLC, a joint venture of Eastman Kodak and Hewlett-Packard to develop digital inkjet photofinishing equipment for mini-labs. Phogenix was short-lived, and dissolved in 2003, just a month after Perez came to Kodak. Many HP employees then left HP and joined Kodak to work on inkjet printing.

2001

Bell & Howell Imaging Acquired

Kodak completed its acquisition in 2001 of the Bell & Howell Company's imaging businesses, consisting mostly of their worldwide electronic equipment service business.

EasyShare System

The company launched the *Kodak EasyShare System,* a new line of digital cameras and docking systems that set the standard for ease of use for digital photography. It made using a digital camera easier for most people, and led to a very successful line of printer docks in 2003.

Kodak Acquires Ofoto

In June, the company acquired Ofoto, Inc., started in 1999 as an online photography service for archiving electronic pictures and enabling online print ordering. Ofoto was renamed *Kodak EasyShare Gallery* in 2005, and renamed *Kodak Gallery* in 2006. The *Gallery* was eventually sold to Shutterfly, Ofoto's main competitor, in 2012.

2002

OLED Prototype Display Shown

Kodak and Sanyo Electric Co. unveiled a prototype 15-inch OLED flat-panel display, the next generation of full-color displays, based on Kodak's patented OLED technology. The OLED technology was getting exciting then, and showed great promise for the future of electronic displays.

Encad Large Format Printer

Kodak bought Encad, Inc. for $25 million, and subsequently the *Encad Large Format Printer* was announced, to compete with the *Iris* inkjet printer (a large-format color inkjet printer that had been introduced in 1985 by Iris Graphics). The *Encad Large Format Printer* made prints up to five feet wide using piezoelectric inkjet technology.

A newer model, the *VinylJet 36* (shown below), came out in 2003. According to the press release, "The digital output provided by *Encad* printers is yet another way in which Kodak is participating within the $385+ billion info-imaging industry, driven by the convergence of images and information technology."[12] These printers were discontinued in 2011.

Fig. 2002.1 *Encad Large Format Printer (EKC)*

End of the Kodak Professional Camera Business

According to Jim McGarvey, late in 2002, Kodak decided to exit the professional camera business due to lack of profit. Nevertheless, a great new Kodak professional camera was announced at Photokina called the *DCS Pro 14n*, with a 14-megapixel resolution and a price of $4,995.

Fig. 2002.2 *The Last Professional DCS Camera (GEH)*

Ironically, my model *Value* was now $V = 3000/4995 = 0.6$, or, approaching one, the highest professional digital camera *Value* so far. Recall we said that the digital age would be here for pictures when $V = 1$! Looks like a bad time to quit the professional camera business in terms of customer value.

2003

EasyShare Printer Dock

In 2003 Kodak introduced the *Kodak EasyShare Printer Dock 6000*, a device that produced durable, borderless 4x6-inch Kodak prints. It employed an ingenious thermal print engine that brought the receiver sheet into and out of the dye-transfer area once for each color. While the process was relatively slow, it was a very small, compact printer that was just right for this application.

It was aimed at user convenience: you simply placed your *EasyShare* camera on top of the dock to make prints without a computer. Many people

Fig. 2003.1
EasyShare Printer Dock 6000 (EKC)

enjoyed taking the unit to other people's homes to make them "instant" prints (OK, prints in 90 seconds). It also came with a software CD for sharing images and/or printing from a computer.

EasyShare Zoom Digital Camera

I won't even try to cover them all, but Kodak introduced many digital cameras in the early 2000s, including the *Kodak EasyShare LS633 Zoom Digital Camera*—the first digital camera to feature an OLED display. The quality of this display was astonishing. Since manufacturing large OLED displays was still a challenge, it made sense for the company to start with small displays, such as the viewfinders on these cameras.

Fig. 2003.2 *EasyShare LS633 Zoom (EKC)*

This camera had a 1533 x 2041-pixel CCD, and sold for only $399, so V = 1533/399 = 3.84. So in 2003, the customer *Value* was greater than one, and on the rise.

2004
Scitex Repurchased by Kodak
In 2004, Kodak repurchased Scitex Digital Printing Inc., by then a leader in high-speed variable data inkjet printing. The new division was renamed Kodak Versamark, Inc. The Scitex Company, located in Dayton, Ohio, had originally been called Diconix when it was part of Kodak in the 1980s, a fact not mentioned in the 2004 press release. Scitex Digital Printing was purchased for $250 million, over three times what it had sold for in 1993. All the while, the Dayton folk had been working on developing high-speed, color, continuous inkjet printing technology as a core competency.

Later, in 2008, the Kodak Research Labs' extensions of the Dayton efforts employing micro-electro-mechanical systems (MEMS) technology were announced at Drupa, the world's largest printing trade show, held every four years in Germany. These innovations were now proudly called *Stream* by Kodak, enabling the *Prosper* line of high-quality digital production printers. These products may be among the few left standing as Kodak emerges from bankruptcy.

NexPress Digital Production Color Press
Kodak acquired Heidelberg's *NexPress* technologies in 2004, originally a joint venture with Heidelberg. The joint venture had been a good fit, with Heidelberg providing experience in sheetfed printing and Kodak supplying the imaging engine. *NexPress* continues to be a provider of high-end, on-demand color printing systems and black-and-white variable-data printing systems. The first *NexPress* digital color press was shown at Drupa 2000. Many versions of this press are available at different price points.

Remote Sensing Systems Business Sold to ITT
"To more closely focus on its growth areas,"[13] Kodak sold its remote sensing systems business, which had served defense and aerospace customers, to ITT Industries. This was the old R&E business that had masterminded

the *Lunar Orbiter* imaging project and the secret reconnaissance satellites. Subsequently, ITT has been doing very well, operating out of the old Elmgrove Plant (now Rochester Technology Park), at the corner of Initiative Drive and Innovation Way (honest!). The company is now called ITT Excelis, with 4,000 employees in the U.S., 1,300 of these in Rochester. Looks like a growth area to me.

Ultralife Sold

That same year, Kodak sold the *Ultralife* lithium battery business, launched in 1986.

Kodak Out of Camera Business!

Kodak also announced that it would stop selling traditional film cameras in Europe and North America, and cut up to 15,000 jobs (about a fifth of its total workforce at the time). Ouch. Kodak out of the traditional camera business? (Kodak continued to sell single-use cameras, however.)

2005

Kodak Polychrome Graphics

Continuing to grow its Graphic Communications business, the company became sole owner of the former joint venture Kodak Polychrome Graphics (KPG)—a leading supplier of products and services to the graphic communications market.

Kodak and Sun Chemical originally established KPG in 1998 with an $817 million investment. For many years, KPG supplied lithographic plates, graphic arts films and papers, proofing products, color management and workflow software, computer-to-plate systems, film processing chemicals and equipment, and related products to the graphic arts industry. Now KPG is part of Kodak's Graphic Communications Group, alongside Creo, Encad, Kodak Document Imaging, Kodak *NexPress*, and Kodak *Versamark*.

My colleague Don Pophal, who became involved in acquisitions later in his career, told me that the KPG acquisition did nothing to help the "loading problem" at Kodak Park, that is, keeping all the equipment running on traditional film products. He felt that this issue of keeping the Park loaded drove a lot of decisions made by Kodak managers over the years relative to emerging digital developments, which they hoped would fail, a tendency I noticed myself at various times.

What's in a Name?
The company's *Ofoto Online Photo Service* changed its name to *Kodak EasyShare Gallery*, and then subsequently to simply *Kodak Gallery*.

CMOS Devices Announced
Kodak announced the availability of the first complementary metal-oxide semiconductor (CMOS) image sensor device to come from its manufacturing alliance with IBM. This new technology led the way to producing lower-cost, high-quality image sensors for electronic cameras and smartphones. More about the effect of smartphones in 2007.

2006
More Digital Cameras
The world's first dual-lens digital still camera, the sleek *Kodak EasyShare V570 Zoom* digital camera, was introduced in 2006. This was followed by the release of two more breakthroughs: the *Kodak EasyShare V610 Dual Lens* digital camera, the world's smallest camera to feature a 10x optical zoom, and the *Kodak EasyShare V705 Dual Lens* digital camera, the world's smallest ultra-wide-angle optical zoom digital camera.

New Kodak Logo
Kodak also updated its brand logo in 2006, "providing a fresh, contemporary look for today's digital world."[14] It was not very popular, however, and, to some, seemed unnecessary. I personally thought that it was unattractive and awkward-looking compared to the previous Kodak logo. Moreover, there were more important things to do.

2007
EasyShare All-in-One Printer
Kodak attacked the high-cost of inkjet printing by introducing the *Kodak EasyShare All-in-One Printer* in 2007. This new printer could produce affordable, crisp documents and Kodak lab-quality photos using low-cost,

Fig. 2007.1
EasyShare All-in-One Inkjet Printer (EKC)

long-lasting, pigment-based inks. The small-particle technology used to make these inks was developed in the Kodak Research Labs.

The first three printer models were the *5000 Series*. The *5100* was a printer, copier, and scanner (hence, *All-in-One*). The *5300* added a 3-inch color display and copy scaling. The *5500* added larger document capability (up to an 8½x14-inch sheet), a fax, and a document feeder.

Astonishingly, even though Perez was still CEO and an enthusiast about the inkjet printers, these Kodak printers never captured enough market share, and Kodak got out of the *All-in-One* inkjet printer business five years later, in 2012.

Health Business Sold to Onex

Kodak completed the sale of its Health Imaging Group (the entire medical imaging business) in 2007 to an affiliate of Onex Corporation of Canada for $2.35 billion. The business is successfully continuing under the name Carestream Health, Inc., employing over 1,100 people in Rochester. As of the first half of 2012, Onex claimed that Carestream proceeds had recovered 106% of its original investment. In 2012, the company launched a new and successful mobile diagnostic X-ray system—*DRX Revolution*.

EasyShare Digital Camera

Also in 2007, Kodak introduced its first camera featuring the company's innovative and low-cost CMOS image sensor technology. The *Kodak EasyShare C513* digital camera offered five megapixels of image capture detail for under $100. This camera had a 2000×2500-pixel sensor, and so V = 2000/100 = 20, about the same *Value* as the early box 35 film cameras. So, it's over, film! (I am totally aware that my *Value* model is simplistic, but, unlike the *Gambit* space programs, this isn't really rocket science. Higher quality for less cost is better, and everyone knows it.)

U.S. Film and Digital Camera Sales

We have seen the digital progression in the above descriptions and in my simple *Value* calculations. I found some numbers here and there and drew a rough graph of film and digital camera sales in the U.S.:

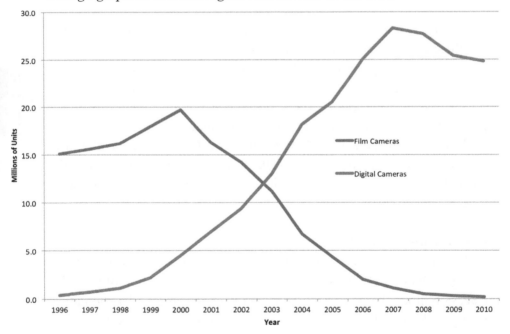

Fig. 2007.2 *Film and Digital Camera Sales in the U.S. (KBP)*

In 1997, Kodak's *DC 120* camera had a *Value* of almost one, and as a result, the digital camera age had started, as hypothesized. In 2003, the *LS633* camera had a *Value* of about 3.8, and film and digital sales were about equal. For film cameras, it was downhill from there. Note the peak for digital cameras around 2007. Why the downward trend after 2007? My guess is that it was due to smartphones, i.e., camera phones!

2008

Mobile Phone Sensor

In 2008, using the new Kodak *TrueSense* CMOS sensor and color filter pattern technology, Kodak introduced the world's first 1.4 micron, 5-megapixel sensor, designed for newly-emerging consumer applications like mobile phones. Imagine, five megapixels in a phone! Apple went ahead and did more than just imagine it; the iPhone 4 currently snuggled in my pocket is in fact a five-megapixel camera. But, oh, right, it's already obsolete.

Stream Inkjet Technology

In the late spring of 2008, Kodak launched more than two dozen new products at Drupa and demonstrated its highly-anticipated *Stream Inkjet Technology*, a continuous inkjet system providing offset-class performance for high-volume commercial printing applications. The technology had been developed over a period of 25 years through the efforts of Mead / Diconix / Dayton Operations / Scitex / Kodak Versamark, if you are tracking all this. Even though the Kodak high-speed inkjet effort was inconsistent over this period due to the sale and repurchase of Diconix, the stream inkjet technology was eventually developed. The resulting *Prosper* printers (see below, 2009) may be part of what is still there after Kodak's bankruptcy is over. I hope so.

First CCD with 50 Megapixels!

The megapixel race was surely on in electronic digital camera marketing. It turns out that the more pixels, the smaller those pixels become, and there is a resultant trade-off with photographic speed (rather like film). But of course, Kodak knew more about that than anybody. Nevertheless, in 2008, Kodak introduced the first CCD image sensor with 50 million pixels, offering unprecedented resolution and detail for professional photography, where photographers take pains to control the light. Amongst the technical folk at Kodak, this CCD provided final, tangible proof that the quality of digital photography was going to rival or exceed the quality of film. For a very long time, that had been considered to be an impossible feat. Not anymore, because now, to convince yourself, all you needed to be able to do was count.

Wireless All-in-One Printer

Wireless versions of Kodak's consumer inkjet printers—the *Kodak ESP 7* and *ESP 9 AiO* (*All-in-One*) printers—were also introduced in 2008. By year's end, more than one million Kodak *AiO* printers had been sold to consumers since the product's 2007 introduction. Not rip-roaring, but a good start.

2009

Stream and Prosper

In 2009, Kodak's breakthrough *Stream* commercial inkjet technology entered the market under the Kodak *Prosper* family name, delivering offset-quality, variable data printing. The first product available was the *Kodak Prosper S10 Imprinting System*, a printhead for use on existing lithographic presses. This technology may be the future of color digital variable-data printing, because it achieves low cost, high speed, and high quality all at once.

Pocket Video Cameras

Kodak continued to grow its line of popular pocket video cameras, introducing the *Kodak Zi8 Pocket* video camera, offering full 1080-pixel, high definition (HD) video capture. Of course, smartphones do this now, too.

Bell & Howell Scanners Acquired

Kodak also acquired the document scanner division of Böwe Bell & Howell, a leading supplier of document scanners to value-added resellers, systems integrators, and end users. It had tried to do this in 2001, but was stalled at that time by anti-trust concerns that are gone now because more companies are in the scanner business.

Kodak Sells OLED Technology

At the same time, Kodak announced that it would sell the assets of its OLED business to a group of LG companies (formerly the merger of Lucky and Goldstar), based in Seoul, South Korea. The Kodak executive quoted in the December 30, 2009 press release explained:

> *As we said earlier this year, OLED is one of the businesses we wanted*

to reposition to maximize Kodak's competitive advantage at the intersection of materials and imaging science. This action is consistent with that strategy. Our OLED intellectual property portfolio is fundamental; however, realizing the full value of this business would have required significant investment.[15]

So, LG made that investment. By 2011, LG Electronics was a $47 billion company that manufactured computer monitors, flash memory, televisions, smartphones, tablets (computers, not pills), mobile phones, DVD players, Blu-ray players, home cinema systems, movie projectors, laptops, and CD and DVD drives. Many scientists today believe that OLED technology is the future of displays, worldwide, because OLED displays are thinner, more energy-efficient, and produce higher quality pictures than LCDs. The Kodak researcher who invented this technology was Dr. Ching Tang, now the Doris Johns Cherry Professor at the University of Rochester, having joined the faculty there in 2006, after 31 years at Kodak.

End of the 2000s

Another decade speeds by, and Kodak's fortunes are in serious retreat. In 2005, Antonio Perez had become Kodak's CEO, taking over from Dan Carp. He was, not surprisingly, initially enthused about inkjet printing, especially after he saw some absolutely lovely small-particle technology at the Kodak Research Labs, but he later abandoned the consumer inkjet business.

Management credibility was continuing to erode, partially because of statements like the following from the 2009 annual report:

While the rate of decline slowed significantly in the fourth quarter of 2009, the level of business activity has not returned to pre-recession levels. During this recessionary period, the Company maintained market-leading positions in large product segments and gained market share with investments into large market categories in need of transformation. The Company ended 2009 with a more efficient cost structure and a strong cash position. The Company believes it is entering 2010 with its most competitive digital portfolio ever, sustainable traction in new markets, value propositions that are embraced by its customers and a leaner cost structure that will help deliver profitable digital growth.[16]

Kodak's worldwide employment decreased from 80,650 in 1999 to 20,250 in 2009; certainly a lower cost structure, but not necessarily "more efficient." The stock was about $5/share, and in two years that would tumble to 20 cents/share.

7

THE TWO THOUSAND TENS

2010

Prosper Press

The first full *Prosper* presses—the monochrome *Kodak Prosper 1000 Press* and the color *Kodak Prosper 5000 Press*—entered the market in 2010. As mentioned earlier, these presses are partly the outgrowth of Diconix that was sold to Scitex and then repurchased by Kodak to become Versamark.

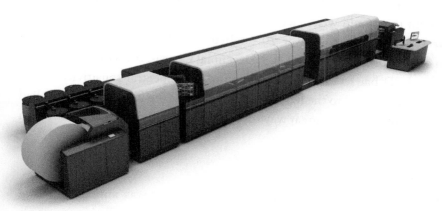

Fig. 2010.1 *Prosper Color 5000 Press (EKC)*

These presses are huge; the little box on the right is a desk!

Another All-in-One Inkjet Printer

The *Kodak ESP Office 6150 All-in-One Inkjet Printer*, with connectivity to print from multiple sources, brought Kodak's revolutionary affordable (yet high-quality) ink to home-based businesses. The *All-in-One* label means that the device is a printer, copier and scanner. These are commonplace today.

2011

Selling More Businesses

"Sharpening its operational focus," according to the press announcement, Kodak sold its Image Sensor Solutions (ISS) business, now operating at Eastman Business Park (formerly Kodak Park) as TrueSense Imaging, Inc. Selling ISS meant that Kodak was now out of the electronic image sensor business. TrueSense Imaging both develops and manufactures imaging sensors for many applications, such as aerial photography, digital cameras, medical imaging and electron microscopy. In 2012, the company received 25 new patents and supplied sensors used in NASA's *Curiosity* Mars rover.

Eastman Kodak Company also completed the sale of certain assets of its microfilm products and equipment business to Eastman Park Micrographics, Inc. This included the *Kodak Archive Writer* and compatible archival microfilm that could write electronic TIFF files to film. The *Kodak Archive Writer* was used to store over 70 terabytes of electronic images scanned from census forms in the 2000 U.S. Decennial Census. It was a slick product, with an LED bar to expose the microfilm. The spectral sensitivity of the microfilm was tuned to the red of the LEDs so that the film could be handled in room light. It is still a great way to archive lots of digital images with high reliability and low cost.

Scanmate Scanner

That same year, the *Kodak Scanmate i920 Scanner* was announced. It is meant for small office spaces and enables businesses of all sizes to easily digitize a versatile range of documents from almost any location. It is a sleek, compact desktop scanner.

Versamark Printing Systems

A new generation of *Kodak Versamark Printing Systems*—with a 40% smaller footprint than competitive systems—was also introduced. The units are for transactional, newsletter, direct mail and newspaper printing, especially where space is at a premium.

Laser Projection Technology
IMAX licensed exclusive rights to Kodak's next-generation *Laser Projection Technology*. This groundbreaking technology enables IMAX screens to deliver high-quality digital content. With continued development, it will be in regular movie houses.

Kodak Spy Satellite Programs De-Classified After Fifty Years
The Kodak spy satellite programs *Gambit*, *Gambit 3* and *Hexagon* were declassified by the National Reconnaissance Office on September 17, 2011. These satellites took high-resolution pictures of Soviet Union missile facilities and other strategic sites from the early 1960s to the mid-1980s. There was a mixture of pride and sadness in the group of Kodak retirees who participated in the events, held at both the Smithsonian National Air and Space Museum's Udvar-Hazy Center in Virginia, and George Eastman House in Rochester, New York. The pride came from being able to finally tell friends and loved ones about the amazing secret projects we had worked on long ago, and the sadness came from being well aware that Kodak was heading for bankruptcy. Here is a picture of me and my beloved boss and mentor from those days, Dr. Don Smith, at the Eastman House event:

Fig. 2011.1 *Don & Brad at George Eastman House (TT)*

As I have indicated before, a complete treatment of these incredible satellite projects would require another book.

The actual reconnaissance satellites, located at the Udvar-Hazy Center, were truly awesome to behold. Here is a *Gambit 1* camera payload, which was denoted KH-7:

Fig. 2011.2 *Gambit 1 Camera Payload (AF)*

The 2011 event was the first time that my wife, Joyce, had ever seen or fully understood what I was actually doing back in the 1960s, and she was astonished. There is a drawing of a *Gambit 1* earlier in this book from 1963, when it started to fly, and you may recall that it was five feet in diameter, 15 feet long and had one recovery vehicle to return the film to earth.

We were busy improving the camera payloads in the 1960s, and started working on *Gambit 3*, which we techies called "Gambit-Cubed" (and sometimes just "G-Cubed") with even bigger and better optics and two recovery vehicles. At the top of the next page is a reproduction of one of the charts made at the Lincoln Plant by one of our designers (they worked on actual drawing boards back then, making all of our charts by hand).

Notice the NRO Approval stamp in the upper left hand corner (otherwise they'd have to shoot me), the SECRET clearance label at the bottom, and the classification category in the lower right, in this case what we called "BYEMAN." (In order to see this drawing in the 1960s, an engineer would have needed a SECRET clearance and also be cleared for BYEMAN documents.) The two recovery vehicles are indicated as RV1 and RV2.

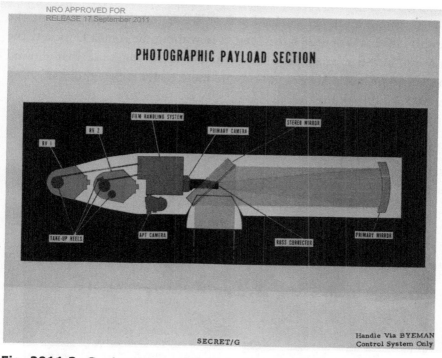

Fig. 2011.3 *Gambit 3 Photographic Payload Section (NRO)*

This drawing shows that *Gambit 3* was indeed bigger than *Gambit 1*:

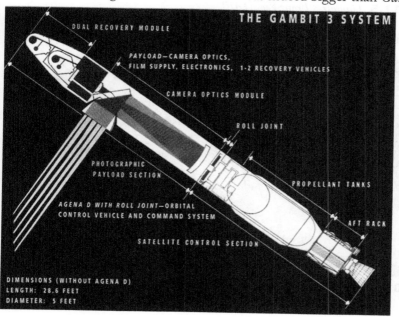

Fig. 2011.4 *Gambit 3 System Drawing (AF)*

Gambit 3's camera payload was still five feet in diameter, but the satellite itself was now 28.6 feet long, even without the Agena D rocket! The drawing on the previous page shows the inventive "roll joint" that allowed the camera to roll sideways and capture more target area without affecting the roll stability of the overall craft.

Gambit 3 was really massive; here is a front photo from Udvar-Hazy:

Fig. 2011.5 *Gambit 3 Front View (AF)*

This rear view shows the orbital control vehicle and command system:

Fig. 2011.6 *Gambit 3 Rear View (AF)*

I was frankly amazed myself to see a full and rather pristine *Gambit 3* up close and personal. You can see it for yourself now at the Air Force Museum in Dayton, Ohio.

The *Gambit 3* rocket was comprised of three parts: the launch vehicle to get the whole thing off the ground from Vandenberg Air Force Base, the second-stage ignition rocket, and the Agena control vehicle, shown in the photograph at the bottom of the previous page.

Here's a drawing of how the overall mission worked:

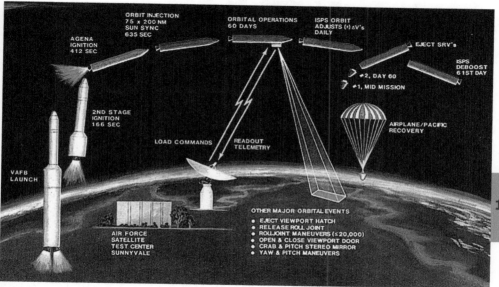

Fig. 2011.7 *Gambit 3 Mission Drawing (AF)*

In order to get the pictures back to earth, the *Gambit 3* dropped the recovery vehicle(s) out of orbit, a parachute would deploy, and some proud and very competent Air Force pilots would snatch the film canister out of the air! No kidding. I was told back then that it was a game among the pilots to see who was best at catching the film. Shame on you if you missed one!

In order to illustrate the point of all this fantastic technology, an actual *Gambit 3* photograph of a missile site in Kazakhstan, which at that time was part of the Soviet Union, is reproduced on the next page.

This photograph shows a booster rocket on the launch pad, so there

Fig. 2011.8
Gambit 3 Photo of a Kazakhstan Missile Site (NRO)

was no doubt that mischief was afoot. All the NRO *Gambit* pictures available to the public, such as the one above, seem to have been purposely degraded, because the actual resolution of the *Gambit* program is still a closely guarded secret, even though the program itself has been formally de-classified. But I know that it was better than the above photo suggests.

Did all this marvelous work stop there? No, it continued, and I was deeply involved in what we called "Valley Pan," working on a design to use two large panning mirrors to get even more coverage and stereo pictures to boot. Due to all the other reconnaissance activity at Kodak (still classified as of this writing), the actual build of this particular camera payload was contracted to Itek, which, we used to joke, meant, "I Took Eastman Kodak." The name of this project became *Hexagon*, shown on the next page.

The *Hexagon* was 10 feet in diameter and 60 feet long—more like a (large) railroad freight car than an *Instamatic*! This actual satellite was also

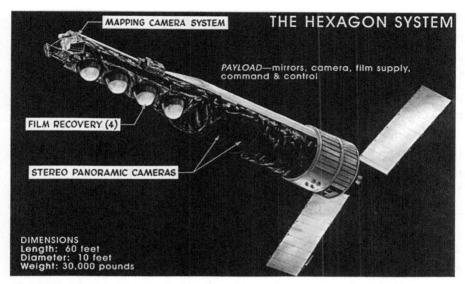

Fig. 2011.9 *The Hexagon System (AF)*

displayed at the Udvar-Hazy event, and it was breathtaking to see, even for us Kodak R&E alumni who were there.

Fig. 2011.10 *Hexagon Payload (AF)*

A significant program called *Bridgehead* (named for the bridge over the Genesee River near Hawkeye) was secretly in place at the Hawkeye Plant during all this time, to process, duplicate and distribute the film from these reconnaissance vehicles. Actually started during the U-2 days, *Bridgehead* supplied President Kennedy with critical imagery during the Cuban Missile Crisis of 1962. From the beginning through the late 1980s, *Bridgehead* processed over 1,500 miles of six-inch film and printed and processed 38,000 miles of duplicate film for intelligence purposes. That's a lot of film.

No doubt you can tell I am still more than a little enthusiastic about this "recce" technology, as we used to call it. Hopefully you can appreciate the incredible photographic technology that Kodak made available in the 1960s and beyond to protect the United States and save us from a nuclear conflict.

2012

Kodak Declares Bankruptcy

Eastman Kodak Company and its U.S. subsidiaries commenced voluntary Chapter 11 business reorganization on January 19, 2012. The CEO said: "We look forward to working with our stakeholders to emerge a lean, world-class, digital imaging and materials science company."[17]

It was a sad day in Rochester and around the world. A lot of us ex-Kodakers called each other to discuss the implications of the decision. But, of course, bankruptcy doesn't necessarily mean the end; there were still new products being developed and announced, like some new *Prosper* presses. However, since that date, Kodak has been selling and getting out of businesses at a rapid pace.

Prosper Presses Expanded

Kodak expanded the *Prosper* platform with the 1000 feet-per-minute, four-color *Prosper 6000XL* press and the 3000 feet-per-minute *Prosper S30 Imprinting System*.

Kodak is Out of the Digital Camera Business

Kodak decided to stop making digital cameras, pocket video cameras, and digital picture frames. Eastman Kodak Company was now officially out of the camera business altogether, which sent shock waves around the world, and was the subject of many newscasts.

Kodak Gallery Sold to Shutterfly
Kodak sold its online photo service, *Kodak Gallery*, to Shutterfly for a paltry $23.8 million. The press release quoted a Kodak executive as saying, "This sale is consistent with our objective of focusing Kodak on a core set of businesses in which we can most profitably leverage our technology and brand strengths, and provides a well-proven mechanism for ensuring that Kodak receives maximum value from these assets."[18] *Kodak Gallery* had about 75 million customers, including myself, all of whose digital photos were transferred over to Shutterfly. Shutterfly won the race; they had always done a better job of promotion than Kodak in my opinion. I once was a *Kodak Gallery* customer; my daughter Holly was with Shutterfly.

ITT Buys Space Computer Corp.
ITT Excelis (created from Kodak's remote sensing business) bought Space Computer Corp., a company that developed signal processing software, to join Excelis's geospatial systems division based in Monroe County, New York. Space Computer Corp. had been based in Los Angeles, had revenues of $14.1 million in 2011, and employed 37 people. This was good news for ITT, but no longer for Kodak.

Kodak Tries to Sell Patents
For several months in 2012, Kodak tried to sell its portfolio of about 1,500 digital imaging patents, but quietly kept delaying the end of the auction. This was because the company (apparently) did not get anywhere near the two-billion-dollar offers it had expected. Kodak was relying on the sale of these patents to help emerge from bankruptcy, and asked for four more months to come up with a reorganization plan.

Kodak Sells More Businesses
In August, Kodak announced that it was selling its still photography businesses, including over 100,000 photo kiosks, and both its professional and consumer still camera film operation. I found it upsetting. About these sales, Mr. Perez was quoted as saying, "We believe they have good long-term prospects."[19] In addition, the document imaging business was put up for sale, including Kodak's excellent line of electronic document scanners, which,

you may recall, had been used to scan 150 million documents for the 2000 U.S. Census.

Kodak Quits Making Desktop Inkjet Printers
One of Mr. Perez's ideas for future Kodak success was to make and sell a new line of desktop inkjet printers, competing with Hewlett-Packard, Epson, Canon, and the like, who were already solidly in the business. These printers were first introduced in 2007, and Kodak attempted to gain acceptance by promoting low-cost ink. Soon after the initial patent deals fell through in the Fall of 2012, Kodak announced that it would cease making desktop inkjet printers but continue to sell ink for the millions of printers already sold, a business that might last three years or so. The headline in the local newspaper read, "Another Kodak plan goes off tracks."[20] A tad snarky, but true enough.

Kodak Sells Digital Imaging Patents
On December 19, 2012, Kodak announced that it had sold its 1,500 digital imaging patents for $525 million. This is only $350 thousand per patent, which is very low, in my opinion. The cost to do the research, write up the invention, get lawyers involved, write up the application and slug it out with the United States Patent and Trademark Office comes out to a lot more than that. This final action on "non-core IP assets" signaled the end of most of the "picture" business for Kodak, perhaps with the exception of commercial printing. There's no turning back from these decisions now.

2013
Innovations Since 2012
The innovations in the picture business just keep on coming, like, for example, *Instagram*, a relatively new online service for sharing photos, linked with social networks like Facebook. All the pictures have a square format, reminiscent of Polaroid cameras and the *Instamatic*, which gives the site a vintage feel. Numerous image filters may be applied to give your photos different looks. Facebook purchased *Instagram* in 2012 for one billion dollars, which made for a little industry buzz at the time.

Here are some more picture innovations I noted in the media in 2013:

January 2013 – Samsung, LG, Sony and Panasonic showed large OLED TV displays at the Consumer Electronics Show. The quality was astonishing, and demonstrated the 4K Ultra High Definition (UHD) TV standard of 3840 x 2160 pixels (8.3 megapixels). The contrast of OLED displays is extremely high compared to today's LCD TVs because since the pixels are point sources of light, when you turn them off they are really, really black.

February 2013 – An MIT student member of the IEEE developed a new processor chip for digital cameras that will allow for better picture quality but less battery drain.

Spring 2013 – ITT Excelis, a company that was started with one of Kodak's satellite businesses, is increasing employment in Rochester by consolidating some ion detector work from Massachusetts, useful in mass spectrometry systems. The devices are also employed in X-ray imaging and scanning electron microscopes.

Spring 2013 – I saw some awesome pictures from a company called Gigapixel.com, which consisted of 100 or more 10-megapixel frames taken with a panning concept. When stitched together with proper software, the result is a picture of over 1,000 megapixels, that is, a gigapixel! You can (digitally) zoom, and zoom!

Summer 2013 – Shutterfly came out with an App for the iPad, *Photo Story*, which makes it easy to create iPad storybooks to share with friends or colleagues. You can add sound, too, and of course, it's easy to order print versions.

If you were to add your favorite innovations that come along here, you'd need a lot of room, depending on how long you keep this book!

8

SO WHY DIDN'T IT WORK?

It's been quite a tale, yet it isn't over, and I know that I haven't covered everything! (In fact, covering everything was not my original objective, because of my presumed finite lifetime.) Nevertheless, there is enough evidence here about electronics and equipment development at Kodak to draw upon for some important learning for future business and technology development students and practitioners. An important question emerges from this study:

What Business Are You In?
The famous management guru Peter Drucker was known for encouraging his clients to ask, "What is our business?"[21] If you think this is a simple-minded question, it's not. After a lot of client discussion and debate, when the group finally landed on some sort of an answer, Drucker liked then to ask, "What should it be?" He was very effective.

As I have said now several times, I personally always thought it was clear that Kodak was in the picture business. But Kodak management could have said, "We are in the chemical business." If that were the case, then when the film business started to wane due to the technological displacement of electronic imaging, the management could have proclaimed, "We are not going to pretend we are in this new electronic imaging business and invest all those billions of dollars in CCDs, electronic cameras, printers and so on, but instead we are going to fire all those electronics types (i.e., sparkies like me) and double down on other applications for our wonderful chemicals!"

In a similar vein, management could have said, "We are in the business of coated consumables," which would, of course, include film, but also many other products, such as photo paper and thermal dye transfer ribbons. How about solar cells coated by the square mile?

Chemistry and coated products were actually part of Fuji's solution, and Fuji is still operating. For example, in early 2012, only one-fifth of Fuji's revenue was in imaging. Fuji has found new uses for its chemical technology in cosmetics, and is applying old film technology to make optical films for LCD flat-panel screens. And in 2000, Fuji bought an additional 25% stake in Fuji Xerox, increasing its earnings.[22]

Kodak also could have said, "We are in the picture business, and if our customers want their pictures in an electronic context rather than on film, we will follow this route." I always felt this way, as I have noted previously. In fact, George Fisher had said essentially that, as I will explain in the next section.

Alternatively, Kodak management could have said, "We are in the film business," but of course, we all now know that that was unsustainable, and many people knew it some time back. Nevertheless, a *lot* of people at Kodak believed strongly that we *were* in the film business, because it was film that made most of the money; they just never explicitly said so. Whenever we did something new in electronic imaging, I was inevitably asked, "Isn't that going to hurt film?" I experienced this many, many times. So I claim, broadly speaking, Kodak could have been more clear about what business it was in. This is my first major conclusion.

Constancy of Purpose

My other major conclusion emerged as I was poring over all the products and technologies I have assembled in this book. There are many examples of what I found, but my favorite, if you can call it that, was the purchase of Mead Imaging, which became Diconix, which was sold to Scitex, and then bought when it was Scitex Digital Printing, Inc. and renamed Versamark. In looking at all of this, I claim that Kodak did not exhibit "constancy of purpose." This phrase kept popping into my head, and I knew I wasn't the first to use it. Sure enough, it is the first point in W. Edwards Deming's famous "Fourteen Points" for business success. Here is what Deming actually wrote:

> *Create constancy of purpose toward improvement of product and service, with the aim to become competitive, stay in business and to provide jobs.*[23]

Read that quote again, please. In my mind, that's what a business is supposed to do. Think about it, in the context of this book or maybe your own business. It's in green for a reason.

I felt that Kodak was in the "picture" business from the beginning. I personally never thought the business was really anything else. If you had asked anybody a few years ago what the Kodak brand stood for, the reply would have been "pictures," without hesitation. Techies liked to use the fancier term "images." Perhaps a normal person would have said "memories," which also means pictures in this context. Along the way, Kodak also made plastic for pop bottles, polyester film base and other things like fibers and so on, but these were not considered Kodak's main products, and were, I believe, only a minor distraction.

However, heavily investing and going into debt to buy Sterling Drug was, to me, the tipping point. Better bean counters than I no doubt will add it all up, but my rough calculation is that the "pill" business adventure cost Kodak in the neighborhood of $10 billion. The venture was not a success largely because Kodak didn't really understand enough about the highly competitive nature of the pharmaceutical business because it was not the company's core competency.

This, of course, leads to one of the underlying ideas behind constancy of purpose and core competency—you need to concentrate and ceaselessly get better and better at what you do best in order to leave the rest of your competition in the dust. Assuming that you could buy the efforts of about ten engineers for a year for about $1 million back in the six-year period of 1988 (when Sterling was bought) to 1994 (when Sterling was sold), then $10 billion could have bought 100,000 engineer-years! You really have to ask yourself what results such a massive army of technical talent could have accomplished if it had been fanatically focused on the picture business.

What Is the Picture Business, Anyway?

I suppose there are some people reading this who might disagree that

Kodak was, or should have been, in the picture business. Just what do I mean by the "picture" business, anyway? For starters, it's a lot more than cameras! It includes all the various aspects of what we often called imaging, besides just cameras, including printing, publishing, displays, print-on-demand, remote sensing, health imaging, document imaging, archiving, scanners, smart phones, Internet photo access, photo labs, materials, imaging software, photo IDs, and movies. There is a lot to the "picture" business! See the drawing below for my chart of these primary aspects of the picture business.

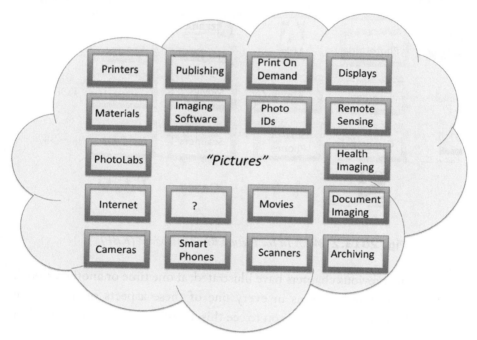

Fig. 2013.1 *What Is the Picture Business? (KBP)*

Now let's see who is active today in all these aspects of the picture business: companies who moved in to take up the slack left by Kodak. Yes, I know, there are many more logos I could put on my second chart (on the next page), but there isn't much room for more logos! I also confess I may have leaned toward some of my personal favorites, but I am trying to make a critical point here. While all businesses have their ups and downs, none of the companies shown here are in bankruptcy as of this writing. Further, they are all enjoying just their little piece of the overall picture business.

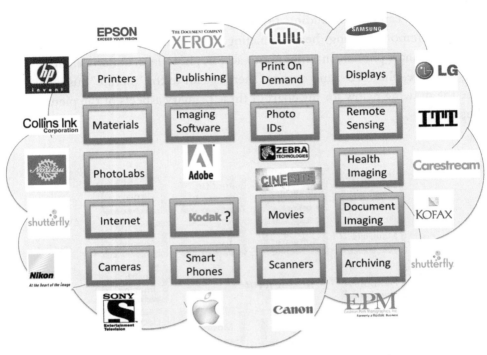

Fig. 2013.2 *Who's In the Picture Business Today? (KBP)*

As the previous chapters have illustrated, at one time or another, Kodak had commanding positions in every one of these aspects of the picture business! I'm not the only person to see this, thank goodness. Validation is a good thing.

About a year after I left Kodak in early 1992, I was called upon to have a meeting with George Fisher, the company's new CEO, because (as I found out later) he kept seeing my name pop up in relation to electronic imaging. I was honored, of course. Our scheduled 45-minute meeting went most of the afternoon. I was pleased at his questions and obvious command of all that I told him, and went home rather excited. I thought that maybe George Fisher was our guy, and he could turn all this around, as for example, when he said in 1997:

It is up to us to envision the future and give direction. In the Kodak case, that direction has been to get back to the fact that we're in the picture business, and in the business of taking pictures further, as our advertisers say. You'll notice I didn't say we're in the film business or the digital imaging business, because our customers and consumers really don't care about film or about digital imaging. They care about their pictures.[24]

Well, this was spot on, as far as I was concerned, and at least I had some good company. I had given a talk to the Kodak board of directors on June 12, 1987 about the promise of electronic imaging, saying that we were in the picture business, not the film business (and a lot more besides, including that traditional photography was getting stale but electronic photography was exciting). Some people thought that I would be fired for saying this. I wasn't, but the response I got was rather tepid—something along the lines of "thanks for sharing."

Four Conclusions

This engineer's meanderings through over fifty years of Kodak's impressive developments in electronics have led me to the following four conclusions about what happened to Kodak:

1. Kodak did not sufficiently clarify what business it was in. Chemicals? Coated consumables? Film? Pictures?

2. Kodak did not exhibit constancy of purpose in the picture business (nor in any of the other possible choices you may prefer).

3. If we assume that constancy of purpose is a necessary (but not sufficient) condition for a technology company to succeed, then Kodak suffered a huge opportunity loss as a result of conclusions #1 and #2 and the resultant lack of internal alignment.

4. Because of the above three conclusions, dozens of other successful companies increased their competency in

electronics and moved in to fill the vacuum, making it impossible for Kodak to regain its lost opportunities.

I think these are the most fundamental issues. The logic makes sense, to me, anyway, and amplifying the internal alignment point, Larry Matteson reminded me of the oft-quoted comment from his business colleague Scott McNealy of Sun Microsystems, who was fond of saying you need "all the wood behind one arrow."

As you know by now, I was always basically an R&D guy, and so believed in continuous technology improvements as a matter of course. This naturally attracted me to a recent book by Gerard J. Tellis, *Unrelenting Innovation*,[25] in which he discusses the "incumbent's curse." It's a good book, and has a section on Kodak called "Decline of an Innovator."

Tellis presents a lot of evidence that this curse leads to three traits that hamper continued innovation: fear of cannibalizing current successful products, being risk averse, and focusing too much on the present. These are much-discussed but excellent points, and I feel that in my experience at Kodak I witnessed the destructive effects of all three of these traits. For example, where X represents a new electronic (non-film) product,

- fear of cannibalizing film led to questions like "Won't X hurt film?"
- risk aversion led to "How will X make as much money as film?"
- short-term focus led to "Won't investing in X lower next year's earnings?"

I have always felt that large companies are a lot like the human body: complex, lots of moving parts, continuous interactions, not easily explained, and so on. Kodak certainly had those characteristics. In particular, though, I think companies, over time, create "antibodies" that attack any foreign objects that threaten the existence of the organism. I recently found that others have made this observation about antibodies, however, it seems apparent that the longer a company is a world leader, the longer these antibodies will have to get trained and become mercilessly efficient, and in the case of Kodak, the time frame was well over a century. Further, I believe that these antibodies were present at all levels of the organization. Steve Sasson suggested to me that this was partly because an organization will tend to act in its own best interest if lacking at least some guidance from the corporate level. This is not to say that I don't appreciate "bottom-up" ideas; I do. However, if a company consists of many organizations all marching to the beat of their own objectives, then a suboptimal corporate system may be the result. As a systems engineer, I have often been fond of saying that, when designing complex systems, "sub-optimization is the name of the devil," and I'd say that that was the case with Kodak.

It seems to be popular these days to criticize Kodak CEOs as being somehow mentally deficient. However, I personally interacted with the last five before Perez (I never knew Perez); Fallon (1972–1983), Chandler (1983–1990), Whitmore (1990–1993), Fisher (1993–2000) and Carp (2000–2005), and I found them to be intelligent, dedicated men who worked hard for the benefit of Kodak. For those who will pursue business cases from here on, I suggest that they might consider the possibility that, because of the powerful antibodies, no single person could have prevented Kodak's eventual demise.

Personal Postscript

Regardless of how or why it happened, which I expect will be the topic of business cases for many years to come, the 2012 Kodak bankruptcy is a sad outcome for a world-renowned company I loved working at for 32 years. Clearly, no one will dispute that Kodak's leadership in the traditional picture business benefitted the world at large, considering the gigantic numbers of precious memories saved for posterity.

We should also note that the world is a better place because of all of the Kodak electronics technology developed over the last fifty years, even though it did not save the day for Kodak. Imagine a world without spy satellites during the Cold War, without the *Lunar Orbiter* and *Apollo* projects, without high-quality office copies, color electronic still cameras, or camera phones, without lithium batteries, compact camcorders, megapixel sensors, jpeg image compression, x-ray medical laser printers, optical discs, single-use cameras, do-it yourself photo kiosks, Internet access to creative photo products, sharp Hubble Space Telescope images, digital restoration of movies, movie special effects, HDTV movies scanned from film, or OLED displays (wait until you see them!). All that technology got a huge boost from Kodak efforts, and we are all better off because of those efforts.

My three grandsons, Kyle, John and David, have heard the name Kodak, mostly from me. However, at present they don't knowingly experience any Kodak products touching their lives. Their children will possibly not even recognize the name Kodak, like Earl and Opal Pickles' grandson Nelson:

Fig. 2013.3 *A 2012 "Pickles" Cartoon (CG)*

ENDNOTES

¹ C. K. Prahalad and Gary Hamel, "The Core Competence of the Corporation," *Harvard Business Review*, May 1990.

² Philip Horzempa, "More Details on NASA's Lunar Reconnaissance Program," *QUEST: The History of Spaceflight Quarterly* 20, no. 1 (2013), 80.

³ The USPTO website: http://www.uspto.gov/about/nmti/recipients/2009.jsp

⁴ Steve Sasson, "A Hand-held Electronic Still Camera and its Playback System," Kodak Apparatus Division Research Laboratory Technical Report, January 12, 1977. Reproduced in *The Camera of the Future* (Germany: DGPh [Deutschen Gesellschaft fur Photographie—Photograph Society of Germany], 2008), 45.

⁵ Harvey Enchin, "Picture this: Camera that has no film," *Gazette* (Montreal), August 25, 1981.

⁶ Christopher K. Veronda, "Kodak Photo ID System Features Breakthrough in Thermal Printing," press release (Rochester, NY: Eastman Kodak Company, February 18, 1987).

⁷ Todd Gustavson, "The First Digital Single-Lens Reflex Cameras—Todd Gustavson talks with James McGarvey," *Image*, Winter 2012/2013, 29.

⁸ Ibid.

⁹ Frank Cost, *Using Photo CD for Desktop Prepress* (Rochester, NY: RIT Research Corporation, 1993), 31.

¹⁰ Emily S. Plishner, "Eastman Chemical spins out of the Kodak Family Portrait," *Chemical Week*, June 23, 1993.

¹¹ Kodak website: http://www.kodak.com/ek/US/en/Our_Company/History_of_Kodak/Milestones_-_chronology/1990-1999.htm

12 Philomena Walsh, "New Wide-Format Printer Enables Direct-to-Vinyl, Outdoor Printing Without Solvents," press release (Rochester, NY: Encad, Inc.—A Kodak Company, April 3, 2003).

13 Kodak website: http://www.kodak.com/ek/US/en/Our_Company/History_of_Kodak/Milestones_-_chronology/2000-2009.htm

14 Ibid.

15 Veronda, "Kodak Completes Sale of OLED Business," press release (Rochester, NY: Eastman Kodak Company, December 30, 2009).

16 Eastman Kodak Company, "2009 Annual Report on Form 10-K and Notice of 2010 Annual Meeting and Proxy Statement," 2010, 4.

17 Veronda, "Eastman Kodak Company and Its U.S. Subsidiaries Commence Voluntary Chapter 11 Business Reorganization," press release (Rochester, NY: Eastman Kodak Company, January 19, 2012).

18 Veronda, "Kodak Enters into Agreement for Proposed Sale of Gallery Photo Services Site to Shutterfly," press release (Rochester, NY: Eastman Kodak Company, March 1, 2012).

19 Matthew Daneman, "Eastman Kodak Exiting Still Photography Business," *USA Today*, August 24, 2012.

20 Daneman, "Another Kodak plan goes off tracks," *Democrat and Chronicle* (Rochester, NY), September 29, 2012.

21 Peter F. Drucker, *The Practice of Management* (New York: Harper & Row, 1954), 49.

22 Kenneth Neil Cukier. "Sharper Focus: How Fujifilm Survived," *Economist*, Schumpeter Business and Management blog, January 18, 2012, http://www.economist.com/blogs/schumpeter/2012/01/how-fujifilm-survived.

23 W. Edwards Deming, *Out of the Crisis* (Boston: MIT Press, 1986), 23.

24 George Fisher, "Kodak Petition—Japanese Trade Barriers" (speech, Boston: Academy of Management, August 11, 1997).

25 Gerard J. Tellis, *Unrelenting Innovation* (California: John Wiley & Sons, 2013), 3.

REFERENCES

Cost, Frank. *Using Photo CD for Desktop Prepress.* Rochester, NY: RIT Research Corporation, 1993.

Cukier, Kenneth Neil. "Sharper Focus: How Fujifilm Survived." *Economist*, Schumpeter Business and Management blog, January 18, 2012, http://www.economist.com/blogs/schumpeter/2012/01/how-fujifilm-survived.

Daneman, Matthew. "Another Kodak plan goes off tracks." *Democrat and Chronicle* (Rochester, NY), September 29, 2012.

———. "Eastman Kodak exiting still photography business." *USA Today*, August 24, 2012.

Deming, W. Edwards. *Out of the Crisis.* Boston: MIT Press, 1986.

Drucker, Peter. *The Practice of Management.* New York: Harper & Row, 1954.

Eastman Kodak Company. "2009 Annual Report on Form 10-K and Notice of 2010 Annual Meeting and Proxy Statement." Rochester, NY: Eastman Kodak Company, 2010.

———. http://www.kodak.com/ek/US/en/Our_Company/History_of_Kodak/Milestones_-_chronology/Milestones-_chronology.htm

Enchin, Harvey. "Picture this: Camera that has no film." *Gazette* (Montreal), August 25, 1981.

Gustavson, Todd. "The First Digital Single-Lens Reflex Cameras—Todd Gustavson talks with James McGarvey." *Image*, Winter 2012/2013.

Fisher, George. "Kodak Petition—Japanese Trade Barriers." Boston: Academy of Management (speech), August 11, 1997.

Horzempa, Philip. "More Details on NASA's Lunar Reconnaissance Program." *QUEST: The History of Spaceflight Quarterly* 20, no. 1 (2013).

McGarvey, Jim. "The DCS Story: 17 Years of Kodak Professional digital camera systems, 1987–2004." June 2004. www.nikonweb.com/files/DCS_Story.pdf.

Paxton, K. Bradley. *Handprint Data Capture in Forms Processing: A Systems Approach*. Rochester, NY: Fossil Press, 2011.

Plishner, Emily S. "Eastman Chemical Spins Out of the Kodak Family Portrait." *Chemical Week*, June 23, 1993.

Prahalad, C. K., and Gary Hamel. "The Core Competence of the Corporation." *Harvard Business Review*, May 1990.

Sasson, Steve. "A Hand-held Electronic Still Camera and its Playback System." Rochester, NY: Kodak Apparatus Division Research Laboratory Technical Report, January 12, 1977. Reproduced in *The Camera of the Future*. Germany: DGPh (Deutschen Gesellschaft fur Photographie—Photograph Society of Germany), 2008.

Tellis, Gerard J. *Unrelenting Innovation*. California: John Wiley & Sons, 2013.

Veronda, Christopher K. "Eastman Kodak Company and Its U.S. Subsidiaries Commence Voluntary Chapter 11 Business Reorganization." Press release, Eastman Kodak Company, January 19, 2012.

———. "Kodak Completes Sale of OLED Business." Press release, Eastman Kodak Company, December 30, 2009.

———. "Kodak Enters into Agreement for Proposed Sale of Gallery Photo Services Site to Shutterfly." Press release, Eastman Kodak Company, March 1, 2012.

———. "Kodak Photo ID System Features Breakthrough in Thermal Printing." Press release, Eastman Kodak Company, February 18, 1987.

Walsh, Philomena. "New Wide-Format Printer Enables Direct-to-Vinyl, Outdoor Printing Without Solvents." Press release, Rochester, NY: Encad, Inc.—A Kodak Company, April 3, 2003.

ILLUSTRATION CREDITS

AF	Courtesy of the National Museum of the U. S. Air Force
CG	Reproduced with permission of the Cartoonist Group.
CHM	Courtesy of the Computer History Museum
DM	Don Maggio Personal Files
EG	Ed Granger Personal Files
EKC	Reproduced with permission of Eastman Kodak Company.
Fortune	*Fortune* ad, April 1963, 171
GEH	Reproduced with permission of George Eastman House International Museum of Photography and Film.
GUSA	Courtesy of Godar USA
KBP	K. B. Paxton Personal Files
NASA	Courtesy of National Aeronautics and Space Administration
NRO	Courtesy of National Reconnaissance Office
Pop Pho	*Popular Photography*, April 1982, 80
RTel	Rochester Telephone Company, 1989 Annual Report, 10
SS	Steve Sasson Personal Files
TSS	Courtesy of Tex's Surplus Sales
TT	Photo by Anthony C. Trotto
U of R	Reproduced with permission of the Department of Rare Books and Special Collections, University of Rochester Libraries.
Wiki	Wikimedia Commons

ACKNOWLEDGEMENTS

I had a fine career at Kodak from 1960 to 1992. During this time I had the privilege of working with many excellent colleagues who taught me a lot about science, electrical engineering, systems engineering, optics, math, management, and much more, including life in general. I was lucky to be there during the best of times.

I have also enjoyed help relative to this book in the form of archival materials, advice, stories and encouragement from many colleagues, business acquaintances and friends, including: Paul Allen, Tony Ateya, Lori Birrell, Mike Clayton, Kathy Connor, Art Cosgrove, Frank Cost, Bob Cosway, Joe DiStefano, Barbara Galasso, Donna Grad, Ed Granger, Todd Gustavson, Keith Hadley, Phil Horzempa, Tim Hughes, Larry Iwan, Pete Lockner, Jerry Magee, Don Maggio, Nancy Martin, Marc Marullo, Larry Matteson, Jim McGarvey, Tom Nutting, Ken Parulski, Jesse Peers, Don Pophal, Majid Rabbani, Richard Reisem, Stu Ring, Akram Sandhu, Steve Sasson, Doug Smith, David Stimson, and Dana Wolcott.

I am especially indebted to my favorite editor, Patty Cost, for her always-relevant, helpful suggestions and editing of the drafts.

The ultimate business case on Kodak has yet to be written, but I look forward to discussing, analyzing, and learning more from those who will work on it in the future, possibly even reading this book along their way.

Fossil Press

We are a private press in Rochester, New York, devoted to producing documentary work that will gain value with the passing of time.

Fossil Press
100 Parkwood Avenue
Rochester, NY 14620–3404

CPSIA information can be obtained at www.ICGtesting.com
Printed in the USA
BVOW10s0545041113

335414BV00001B/1/P